MW00463143

SPACE FORCE!

A QUIRKY AND OPINIONATED LOOK AT AMERICA'S NEWEST MILITARY SERVICE

TAYLOR DINERMAN

PERMUTED
PRESS

A PERMUTED PRESS BOOK
ISBN: 978-1-68261-983-4
ISBN (eBook): 978-1-68261-984-1

Cover art by Cody Corcoran

PERMUTED
PRESS

Permuted Press, LLC
New York • Nashville
permutedpress.com

Published in the United States of America
1 2 3 4 5 6 7 8 9 10

In Loving Memory of My Father

Sam Dinerman
Athlete, Soldier, Salesman, Bon Vivant

CONTENTS

PART FOUR
The Future

ACKNOWLEDGMENTS

First of all, I want to thank the late Brigadier General Robert (Bob) C. Richardson III, USAF ret.; more than anyone, he was my mentor in military space affairs, missile defense, and the mysteries of the Pentagon's bureaucracy. I also want to thank the late Milnor Roberts as well as Hank Cooper from the High Frontier organization.

I also want to thank Coyote Smith, Chris Stone, Marc Dinnerstein, Brent Ziarnick, Christopher Bosquillion, and my editor at the *Wall Street Journal*, James Taranto. You guys are the greatest.

I cannot forget to thank my friends in the space industry, Dale Amon, currently CEO of Immortal Data; Ivan Bekey; Lukas Viglietti; and Ty McCoy of the Space Transportation Association.

For their help and support over the years, I am grateful to Buzz Aldrin, Howard Bloom, Jeff Foust, Peter Garretson, Brandi Gaudet, Ken Hodgkins, Jack Fowler, Gordon Pennington, Peter Huessy, Nina Rosenwald, Jess Sponable, and Jerry Tubergen.

And Always, Sandy.

PREFACE

I am not a space expert. I have, however, been observing "space" since my childhood days in Washington DC, where the media environment was saturated with Project Mercury, Walt Disney's Tomorrowland, and the Cuban Missile Crisis. In 1969, at a low point in American history, I watched—on Swiss TV along with my father—the Apollo 11 Moon landing. If my country could do that, it challenged, in my young mind, the tidal waves of anti-American propaganda I saw and read almost every day in the European media.

I've been writing about space, strategy, nuclear weapons, and—sometimes—cultural issues since the early 1980s. In particular I've long been interested in missile defense. The idea that our survival would be based on nothing more or less than the threat of nuclear retaliation never seemed, to me, quite rational.

I am also convinced that, as Tsiolkovsky put it, "The Earth is the cradle of humanity, but one does not stay in the cradle forever." The human race will move out into the solar system and, probably, someday, beyond it. As this happens, we will not cease to be human, with all the virtues and all the flaws inherent in our species.

The United States Space Force, therefore, has a noble mission: to protect America, to support our allies, and to support our nation's interests as we, along with other nations, move out into the little patch of the universe that is our home system. As an institution it has gotten off to a shaky start, but I have great confidence that someday soon it will find the right leadership and will eventually be emancipated from the Department of the Air Force.

At that point it can begin to truly serve the great cause of creating a spacefaring civilization—as it was always meant to.

INTRODUCTION

It will be years before competent historians gain access to the relevant archives and the perspective to tell the tale of how we got to the Trump administration's Space Force proposals, then to a new military service in 2019.

In 1983, in a book that went to press a few months before president Ronald Reagan gave his famous "Star Wars" speech, Thomas Karas, supported by the Federation of American Scientists, published a book called *The New High Ground: Strategies and Weapons of Space-Age War*. He noted that some of the Air Force officers he called "spacemen" were happy to see the administration creating a Space Command: "Other spacemen see an eventual need for a US Space Force. They believe space is as different a medium for military operations from the air as the air is from the sea, or the sea from the land." What seemed ridiculous in 1983 is commonplace in 2021. The physical nature of space has not changed, and while technology has improved, the essential functions of military space, communications, navigation, early warning, intelligence gathering, and anti-satellite operations are the same in 2021 as they were in 1983. So what changed?

In 1995 at a symposium on Air Force Space Operations, USAF historian George W. Bradley III asked, "Does the command [Air Force Space Command], the birth of which was aptly described by [General Earl] Van Inwegen, represent the forerunner of a true space force? It took three decades to go from a Signal Corps aerial component to an Air Force, and it may take that long again to go to a Space Force. Whether we get there or not may not be so much a question, but simply a matter of time."[1] Twenty-four years later it would seem we are on the verge of having a United States Space Force, but not yet a full-blown Department of Space.

The US Space Force idea has been around for decades, in both science fiction and in the minds of people who attempt to seriously consider what our nation needs in order to deter future wars and, if necessary, to fight and win them. In early 2001, before the 9/11 attack, it seemed as if Donald Rumsfeld was ready to propose something very like the "Space Corps" idea (a new service inside the Air Force comparable to the way the Marines are part of the Navy Department.) As one USAF space officer put it, "Finally we have someone in the SecDef's office who loves us."

After the attack, Rumsfeld had other and more urgent work to do. While many people in the defense establishment still spoke in terms of "transformation," the major reforms that had been expected never happened. A typical example of this sort of thing was the failed attempt to build a "Transformational Satellite Communications System."

But the Space Force idea refused to die. The post-9/11 Global War on Terror required massive communications capability. Satellites that were used to control drones over Afghanistan from bases in Nevada were soon overwhelmed. And while the Air Force and its contractors found some ingenious ways to work around the limitations of existing systems, it was obvious that new ways of doing business were required.

Yet Congress did not want to create an extra four-star position devoted to homeland defense; this seems to have forced Rumsfeld and the Bush administration to disestablish US Space Command and place its functions under US Strategic Command in order to allow for the creation of Northern Command, whose job is homeland security. Over time, for military space operations, this proved even less satisfactory than what had existed before.

Strategic Command, whose primary emphasis is nuclear deterrence, failed to give the space mission the support it needed. Indeed, over the years, the Omaha-based organization has proved less than adequate at most of its functions, but that is probably due more to failures of the political leadership in Washington than to anything else.

Under the US Constitution, the interplay between the president as commander in chief, the Congress with its power of the purse, and the military, is inherently difficult. All democracies have this problem to one extent or another, but in America we often make things harder on the military than even a strict interpretation of our founding document would require.

One way we do this is by electing some truly stupid people to Congress. One prominent New York congresswoman is supposed to have replied to some constituents who were urging her to support a human trip to Mars with the words, "Haven't we been there already?"

In any case, neither George W. Bush nor Barack Obama were willing to take the time and make the effort to think about what an effective US space policy would look like. It was left to a small band of congressmen and senators to try and elaborate what reforms were needed. On the Senate side, Jon Kyl (R-AZ), while he was in office, did the heavy lifting, but in the end, it was Mike Rogers (R-AL) and his partner Jim Cooper (D-TN) who provided the decisive leadership.

In 2017, they introduced a bill that would have created a "Space Corps" inside the Department of the Air Force, supposedly an analog to the way the Marine Corps is part of the Department of the Navy. The proposal was opposed by the powers that be inside the Defense Department, but inside what can be referred to as the military space community, the reaction was, "At last!!!"

Rogers's bill easily passed in the House with bipartisan support. Sources indicate that Rogers cut a deal with Senator John McCain to support McCain's desire for a new Cyber Command—independent of the National Security Agency—in exchange for Senate support for the Space Corps.

In any case, according to sources, Senator McCain reneged on his promise. It may be that he cut some sort of deal with Air Force leadership, or maybe he was just being his usual obstreperous self. In any case, it looked as if the Space Corps idea was dead. But no one was counting on president Donald Trump to act decisively. Aside from his promise (fulfilled) to reestablish the White House Space Council under vice president Mike Pence, many specialized observers believed that he had even less interest in space than his predecessors.

We were wrong.

One can guess that space exploration and US military superiority in space are integral to what Trump sees as "American Greatness." In this he is not alone. For many baby boomers, the Moon landing in 1969 redeemed an America which looked (on TV anyway) as if it were coming apart. Meanwhile, *Star Trek* (with its quintessentially American, optimistic vision of the future) and the first *Star Wars* movie (whose audiences refused to get the subtext that it was the US that was the evil empire and not Leonid Brezhnev's USSR) just added to the mystique of "space."

Perhaps, then, Trump's imagination did not need much stimulating after the Space Corps idea was shot down. In an April 8, 2019, interview in *Defense News*, Representative Cooper claimed that, "Well, the president obviously hijacked our proposal and exaggerated it." There is, however, a big difference between a president's ability to publicize an initiative and that of a group of congressmen, no matter how well thought-out their proposal may be. Indeed, when Trump was president, no one in the legislature could even begin to match his PR firepower.

According to knowledgeable sources, he'd been following the debate and also was being briefed on the almost-daily cyberattacks on US satellites, as well as on Russian and Chinese anti-satellite weapons programs. The antipathy between McCain and Trump probably pushed the president toward a Space Force decision on the principal that "whatever he hates I am prepared to like."

Two of the most significant outside influences on Trump's decision were former House Speaker Newt Gingrich and the then-House majority leader (now minority leader) Kevin McCarthy (R-CA). Along with a few knowledgeable people in Washington, they knew that the Air Force had neglected space.

Ever since the Air Force was formed in 1947 along with the Department of Defense, the reform of US military structures has periodically emerged to trouble Washington's political establishment. In 1949 there was the "Revolt of the Admirals"; in the early 1960s Robert McNamara tried to reform the department along more or less civilian corporate lines, especially with his ill-judged emphasis on "cost effectiveness." In the 1980s, a military reform movement succeeded in making a few minor changes to the way the Pentagon handled procurement. None of these reformers ever thought to propose a whole new service.

Trump may have been surprised at the positive reaction he got from his supporters when he first publicly came out in favor of the Space Force. The loud cheers and chants of "Space Force, Space Force" at his rallies whenever he mentioned the idea were a sign that his most fervent supporters "got it," even if the bureaucracy in Washington did not.

The opposition fell into two distinct camps. First of all, there were (and still are) the anti- American "peace" camp. These folks are mostly old-style leftists who've devoted their lives to weakening American military power. Their arguments against "militarizing space" or keeping space as a sanctuary only appeal to the uninformed or to left-wing extremists. The other group that opposed a space corps are both more credible and more influential. These are people in and out of uniform who believe that the status quo is acceptable and that there is little about military space, aside from more funding, that needs to be fixed.

This second group makes the case that the new service will be just another bureaucracy—that the Space Force will bring nothing to the US military as a whole that is not already being done. Yet the Air Force's problem with space system procurement persists. Satellites and their ground-control elements are difficult to build and operate at the best of times, but the long-standing pattern of overpromising, followed by delays, cost overruns, and "restructuring," is one of the primary reasons why a new service is needed.

One example of this frustrating pattern is the SBIRS GEO satellites. In spite of delays and cost overruns, these early warning, heat-detecting satellites (that sit in stable geosynchronous orbit roughly 26,000 miles from Earth) have been providing the military and the intelligence community with whole new-level, quality information about what is happening on our planet.

However, due to their high cost, the seventh and eighth satellites in the series have been canceled in favor of a new program that will use a larger number of smaller spacecraft that may or may not provide the same quality of information. Programmatic changes such as these are risky and may add to the expense of the future system. The Space Force may be able to reduce the number of such disruptions by ensuring that when large and expensive programs begin, they are state of the art and are properly funded. This is not simply a function of better trained procurement professionals but is also a function of senior leadership. The top generals of the Space Force will have to be able to say no to seemingly promising proposals unless they are convinced the program can be executed within a reasonable time frame and can keep to something that remotely resembles the original cost estimates.

Finally, on June 18, 2018, Trump made it clear: "I'm hereby directing the Department of Defense and Pentagon to immediately begin the process necessary to establish a space force as the sixth branch of the armed forces. That's a big statement. We are going to have the Air Force and we are going to have the Space Force—separate but equal. It's going to be something. So important. General Dunford, if you would carry out that assignment, I would be greatly honored also. Where's General Dunford? General? Got it?"

To which General Joseph Dunford, USMC, the chairman of the Joint Chiefs, replied, "Got it!"

The process thus began, and Dunford was supposed to be personally in charge. Which brings up two points: Was a Marine the right guy to give the job to? And how could the Trump administration prevent the Air Force establishment from sabotaging the effort?

The Marine Corps is a small service; traditionally, it is the least bureaucratic of all the services and the most combat-oriented. Its role is not to control territory, or to rule the sky or the waves, or to deter nuclear war—its reason for being is to get close to and destroy the enemy.

There are a few Marines who "get" space. Over the years, they've tried and—for the most part—failed to push the Corps toward developing a space role. The idea of a Marine global mobility vehicle that would launch like a rocket, land vertically, and deliver a squad of Marines to any point on Earth within an hour or an hour and a half was proposed and rejected during the George W. Bush administration. (The idea may be revived if Jeff Bezos's Blue Origin makes good on its plan for "point-to-point" passenger rocket services.)

In any case, after months of political and bureaucratic haggling, the Defense Department came up with a plan that would indeed create a Space Force—essentially a modified version of the 2017 Space Corps idea. This plan was the beginning of the Space Force story.

On February 19, 2019, Trump signed his fourth Space Policy Directive aiming to set up the Space Force as a separate service inside the Department of the Air Force. The plan went to Congress where, after some of the usual rhetorical back-and-forth, it proved acceptable since it dealt with almost all of the serious objections raised in both the House and Senate. It would not require much more money; indeed, reports at the time indicated that only $100 million in extra funding would be required.

But, in the long term, this looked more like a first step toward Trump's goal of a fully independent Space Force, rather than a permanent fixture within the Department of the Air Force.

This first Space Force proposal was carefully tailored to respond to Congressional concerns. Unlike other Trump initiatives, it carefully respected legislative prerogatives. It even bent over backwards to give Congress a say in personnel decisions that would normally come under the president's purview as commander in chief.

The proposal did not provide for a Space Force Academy. This may or may not be important; the Navy and Marines have done reasonably well sharing the Annapolis graduates. Sadly, all of America's military academies have developed serious quality problems over the years. A Space Force Academy might be a way to avoid having the USSF contaminated by the dysfunctional culture we see at the USAFA.

The rest of the Space Force proposal looked, at first glance, to be well suited to a start-up service. It did not demand that the National Reconnaissance Office (NRO) be placed under the USSF. It did, however, require that the NRO work closely with the new service to deliver better and more timely information to the warfighter and to the other people in the government who need it.

In an article published in *Space News* on March 2, 2019, one anonymous official was quoted as saying, "We're trying to establish a unique culture, with special training, promotions, doctrine ..."[2] This was exactly the right goal for 2019 and indeed is for most of the next decade. Unlike the Air Force in 1947, the Space Force is not emerging after a major war. Back then there was a huge number of men, and a very few women, who were trained and ready to fit into the new service. The Space Force will initially be manned by roughly 15,000 personnel taken mostly from the Air Force.

Selling this proposal on Capitol Hill proved easier than expected. In the age of Trump, few bipartisan initiatives ever seemed to get passed (criminal reforms were one exception). Sadly, as the 2020 election got closer, we saw more and more grandstanding and nastiness on all sides. Yet, amazingly, the Space Force battle was won and the new service was established by law in the 2020 Defense Authorization Bill.

Meanwhile, China and Russia were rapidly improving and expanding their space warfare capabilities. As the months and years continue to pass, the need for a US Space Force becomes more and more evident.

NOTES ON SPACE

A quick note on the nature of the space domain.

It is generally accepted that referring to "space geography" is a *faux pas*, since the prefix "geo" refers to Earth, and the one thing space is not is Earth. A similar objection applies if one tries to describe "the terrain of space." The US military has, for the moment, decided to call it a "domain," which, while not as misleading as "geography" or "terrain," still implies that the space domain is somehow a house, home, or domestic environment.

Space is an environment even more dominated by the force of gravity than is our Earth. To get to low Earth orbit (LEO) you need to achieve a speed of 17,500 miles per hour—roughly Mach 25. Taking off from the surface of the Earth, a launch vehicle needs roughly 90 percent of its total weight to be fuel; in the case of a liquid-fueled rocket, that fuel will be liquid oxygen and something like kerosene, liquid hydrogen, or, in the case of the Nazi V-2, alcohol.

This creates what is termed the mass fraction problem. When a space launch vehicle is being built, the designers have only 10 percent of the total weight for structure, propulsion, and payload. This is why rockets are big and satellites are small. Until very recently, rockets were designed to fly in stages. As the fuel in each stage was expended, the stage and all its hardware were thrown away. This was referred to as "feeding very expensive bits of titanium to the fish." Advocates for a new type of launcher often ask the question, "How much would it cost to fly from New York to Paris if you had to throw away 90 percent of the airplane every time you went?"

To get around this, various reusable rockets were proposed. It was not, however, until the Strategic Defense Initiative Organization (SDIO), under the direction of Hank Cooper and using the design philosophy of legendary rocket engineer Max Hunter, that the DC-X experimental vehicle program first flew in 1993. The DC-X, also called the Delta Clipper or Delta Graham (after Dan Graham, who headed the High Frontier organization and helped lead the lobbying effort for the project), showed what could be done—that the idea of reusable launch vehicles (RLVs) could be taken seriously.

In the 1990s and later, NASA and the Department of Defense (DoD) tried and failed to design and build RLVs, but it was not until Elon Musk began having the first stage of his Falcon 9 and Falcon Heavy rockets land, either on barges or on land, that the RLV concept was accepted as practical. Meanwhile, Jeff Bezos's Blue Origin, which hired some of the original DC-X team, has been following a step-by-step process that includes reflying its New Shepard vehicle on suborbital test flights and designing the giant New Glenn launcher.

Musk is working on a fully reusable Starship, which he hopes will provide a way to economically fly to the Moon, Mars, and to LEO. This could go into service as early as 2024.

As of now—and for the foreseeable future—LEO has the most human activity. LEO is generally defined as 100 to 1,000 miles (160 to 1,600 kilometers) above the Earth. The International Space Station is in LEO at 248 miles altitude (on average). Most spy satellites and commercial and scientific remote-sensing satellites are located in LEO as well, and an increasing number of commercial communications constellations are located in LEO. For military purposes, dominating LEO is the key to dominating the rest of the Earth-Moon system.

Medium Earth orbits (MEOs) are defined as being between 1,000 to 12,000 miles (1,600 to 19,300 kilometers). The Global Positioning System (GPS) satellites and other Positioning Navigation and Timing (PNT) spacecraft travel in these orbits.

At 22,236 miles (35,786 kilometers) is geosynchronous equatorial orbit (GEO), sometimes called geostationary orbit, where the big expensive communications and early warning satellites dwell. Slots in GEO are a valuable commodity, and their distribution is controlled by the International Telecommunication Union based in Geneva.

Highly elliptical orbits (HEOs, sometimes called Molniya orbits) are oval-shaped orbits in which spacecraft travel between 660 and 24,000 miles (1,060 to 38,624 kilometers) above Earth's surface. These are used for specialized communications and observation satellites. The Russians have used this type of orbit for communications satellites which serve their requirements in the Arctic. The US uses them for at least a pair of early warning sensors. These obits are also useful for electronic intelligence-gathering satellites.

There are five Lagrange points in the Earth-Moon system, named for the eighteenth-century Franco-Italian mathematician who postulated their existence. Simply put, these can be defined as the orbital spots where the Earth's and the Moon's gravity field cancel each other out. The points labeled L4 and L5 are potentially useful for large-scale activities; the others are wildly unstable but can be used as sites for scientific satellites such as NASA's successful solar physics spacecraft SOHO. In the 1970s, Gerald O'Neil and others also proposed that a space colony be established at L5 to manufacture space solar-power satellites out of materials mined on the Moon.

There are other Lagrange points in the solar system created by the gravitational fields of the sun, the planets, and other objects.

Above most of the Earth is a pair of radiation belts, called the Van Allen belts after the US scientist whose instrument, launched on America's first satellite, discovered them. They protect Earth from most, but not all, harmful solar radiation. These belts do not cover the polar regions, and blasts of solar radiation spill over the radiation belts in the far north and far south, thus creating the phenomena called the Northern Lights or the Aurora Borealis.

The Moon travels around the Earth in an elliptical orbit at an average distance of 238,000 miles (384,000 kilometers). The Moon itself is 2,160 miles in diameter. Compared to most other moons in our solar system, it's pretty big. As a site for building a settlement, it's not ideal; any humans on the surface would be subject to the full force of solar radiation, not to mention cosmic rays. However, it should be possible to bury habitats under the lunar soil (regolith). What makes the Moon truly attractive is its proximity to Earth and its low gravitational field. Blasting off from the Moon takes far less energy than doing so from Earth.

This is referred to as the gravity well.

For most of the rest of this century the only other places of economic and military interest are Mars and the asteroids. Back in the days of the French Empire, they used to refer to southern Chad as "*Le Tchad utile*" (the useful part of Chad), implying that the rest of the country was useless. In the long term, the useless part of Chad turned out to be more desirable than the French authorities thought it was. In time, parts of the solar system that are currently considered economically useless could prove valuable, but probably not until the twenty-second century.[3]

PART ONE

SHAPING THE FORCE

CHAPTER 1

THANKS, SENATOR

*Thanks Senator. You did your job, you asked the right
questions, and you came to the right conclusion.*

—Taylor Dinerman

When Senator Jim Inhofe (R-OK), chairman of the Senate
Committee on Armed Services, announced that he was
reluctant to support President Trump's proposal for a Space Force, he
said, "If it's not broke, don't fix it." Unfortunately, America's military
space enterprise *was* "broke" and had been broken for decades; a radi-
cal change was needed. Congressman Adam Smith (D-WA), who probably should have known better, claimed that a Space Force would be
"wasteful." Why would a Space Force be more wasteful than the Space
Corps idea he voted for in 2017, which was essentially what Trump
proposed in 2019? In the end, fortunately, both men voted in favor of
the Space Force.

Senator Inhofe has his virtues—and no doubt Congressman Smith
does too; he certainly deserves unstinted praise for the tactful and scru-
pulous way he handled his role as temporary head of the Armed Services
Committee during the last months of John McCain's life. Inhofe's vote
in favor of the USSF was a big step toward solving a persistent and
dangerous threat to America's position at a relatively small cost. If
America were to lose the use of its military space systems, the result

would be catastrophic. The three critical areas are communications, navigation (i.e., Positioning, Navigation and Timing—PNT), and various forms of surveillance, sensing, and intelligence gathering (including optical imagery, radar imagery, infrared imagery for early warning, and electronic intelligence collection grouped under the acronym ISR). The loss of any of these capabilities would require the military to use less capable alternative systems and methods, which are vulnerable to countermeasures. Space has finally been publicly acknowledged as a "contested domain" after decades of denial and delusion on the part of the US government (not to mention some Washington think tanks). The lengths to which some beltway types went to avoid dealing with the reality of space warfare is amazing.

Right from the start of the Space Age, the Air Force staked out its claims to control military space operations. In 1958, for example, then-USAF chief of staff Thomas D. White wrote, "In discussing air and space it should be recognized that there is no division, per se, between the two. For all practical purposes air and space merge, forming a continuous and indivisible field of operations. Just as in the past when our capability to control the air permitted our freedom of movement on the land and seas beneath, so in the future will the capability to control space permit our freedom of movement on the surface of the earth and through the earth's atmosphere." It has long been obvious that space operations are nothing like air operations; the physics alone makes it impossible to use air tactics or operational concepts in space. While intercontinental ballistic missiles (ICBMs) were and still are a natural part of the Air Force's nuclear deterrent mission, satellites and space operations are very different animals from air combat.

One sign of just how bad things were is that every few years the leadership at the DoD felt the need to reorganize the way military space was managed. In 2009, for example, the incoming Obama administration abolished the National Security Space Office (NSSO), which was the only part of the US government trying to tie together all of the space activities being carried out by the Pentagon, NASA, the intelligence community, the National Oceanic and Atmospheric Administration (NOAA), and the rest, not to mention its role working with what was then the emerging "New Space" industry. They also canceled the high-capacity Transformational Satellite Communications System program (TSAT) and tried to abolish the small Operationally Responsive Space Office, which has been doing such extraordinary work that it is now one of the inspirations for the new Space Development Agency (SDA). These reorganizations should be a bright, shining sign that the US military has not figured out how to manage its space assets. The problem was summarized in the so-called "Rumsfeld Report" published in January 2001: "Many believe the Air Force treats space solely as a supporting capability that enhances the primary mission of the Air Force to conduct offensive and defensive air operations. Despite official doctrine that calls for the integration of space and air capabilities, the Air Force does not treat the two equally."[4]

In June 2010, the space industry was revolutionized by the launch of the first SpaceX Falcon 9 rocket. Elon Musk's firm was the first nontraditional enterprise to successfully enter the launch business. In the 1980s, Orbital Sciences had tried to start a similar process of change, but the company eventually evolved into a more traditional aerospace contractor and were absorbed by the Northrop Grumman conglomerate. The Falcon 9, which even in its early versions cost a lot less than its competitors, was soon able to earn the lion's share of the worldwide commercial launch business, leaving the French (European) Ariane launchers and the Arianespace version of Russia's Soyuz with most of the rest, as well as with the captive European market. In the last decade, Musk's firm has continually lowered its costs, thanks to both reusability and constantly improving manufacturing techniques. For the US military, this has led to lower costs and more opportunities to fly satellites of all kinds. Musk has given American spacepower an unexpected boost; so far, though, the DoD has only made a modest start at taking advantage of this.

The Air Force and the intelligence community have a long-standing preference for very expensive and very vulnerable "exquisite" space systems. There is no question that these systems are amazingly good at what they are designed to do. The Space-Based Infrared System (SBIRS) early warning satellites are proving to be extremely valuable; their role in helping the US military understand exactly what the North Koreans were up to with their missile development programs gave President Trump the information he needed to decide on his denuclearization strategy. Their contribution to our knowledge about the Syrian Civil War has not yet been fully explained to the US public, but one suspects that it has been considerable. Our large, intelligence-gathering imagery satellites, often referred to by their old code names "Keyhole" for the optical ones and "Lacrosse" for the radar imaging ones, reportedly can detect very, very small objects and give enough raw data to America's imagery interpreters for them to do an amazingly good job. No other nation on Earth can match the US in this field, but the cost in time and money to achieve this capability is very, very high. Even worse, neither the USAF nor the NRO has been able to—or even allowed to—put active defense systems on these multibillion-dollar satellites.

The belief that space can be made into a "sanctuary" where active or kinetic weapons are never used is the kind of damaging fantasy typical of the American "arms control" community, which has, over the years, promoted ideas such as the Nuclear Freeze and putting nuclear weapons under our cities and giving the Soviets control over them, all the while expecting Moscow to allow us to do the same with its cities. The US must be prepared to ignore this concept. Not only have the Chinese and Russians made extensive preparations to fight in space with almost every type of weapon, but the refusal of the government in Washington to take the threat seriously has weakened our overall deterrent capability. The recent inability of the Defense Department's senior leaders to think clearly about space and to enunciate a clear policy has left America's military space operators without the guidance they need to prepare to actually fight in space. In turn, this has led to even more confusion and waste than usual.

Todd Harrison, a space expert at the respected middle-of-the-road think tank CSIS, pointed out that the Navy's MUOS broadband communications satellites are in orbit, but that they are useless since the Army failed to find the funds to build the ground segment. This has international repercussions, since the MUOS is being built as a complex partnership with Australia.

Another reason why we need the United States Space Force is that given the decades-long institutional resistance to space-based missile defense, only a new organization whose mission is centered around building up American spacepower can be counted on to develop such a system in a reasonable time frame without having to constantly worry about whose bureaucratic toes get stepped on. The Brilliant Pebbles program that was canceled in 1993 by the Clinton administration would have given this country a serious multi-layered missile defense system instead of the bare-bones one we now have. The FY 2019 Defense Appropriations Bill provided money to begin serious work on a space-based missile tracking system that will complement the current early warning satellites and the complex of surface radars that our missile defense system now relies on to detect and trace missile launches. The technology used for these tracking satellites can be applied to active interceptor systems without too much trouble.

In 2002, in the aftermath of the 9/11 attack, the Pentagon abolished the US Space Command and moved its functions to the US Strategic Command, where "space" was just one of a variety of the command's responsibilities. This was done because a new home defense command (Northcom) was deemed necessary, and Secretary Rumsfeld, in spite of his known "pro-space" proclivities, was reluctant to try and convince Congress to give permission for a new four-star position.

The Air Force's effort to cultivate a cadre of space warfare experts has been bouncing between various commands with little focus or accountability (see chapter 2). Officers who devote themselves to building American spacepower tend not to get promoted, even if they also have the traditional flying background. In 2010, the National Security Space Office was abolished and its expertise scattered. This is a good example of the way military space people get treated by the Air Force and the Defense Department as a whole.

For decades, military space has been treated like an unloved and unwanted stepchild by many of our military and political leaders. A few visionary generals, junior officers, and politicians have tried and failed to fix things. More than a few officers have seen their careers cut short because they persisted in trying to give America the spacepower it needs and deserves.

There are two things that a Space Force would give us—and they are two things that are sadly lacking in the current structure: 1) a realistic space warfighting doctrine that focuses on US space dominance and protects the US ability to use its space assets all the time in war, peace, and the grey area in between,[5] and 2) a cadre of military space leaders who understand both the space environment and space technology and what it can and cannot do. The Air Force programs aimed at buying and operating space systems have been plagued by a lack of professional expertise and a lack of expert outside supervision. The delays and cost overruns associated with the SBIRS early warning satellite program are just one of many examples of this problem.

There is also the question of the way space systems funding has been treated. Money, as we all know, is fungible. A good example of this is the way that funds sent to the Palestinian Authority for humanitarian purposes allow the PA to use other funds for terrorist support. For years now it has been claimed that the Air Force was shifting funds out of space programs and into aviation programs that the service thought were more important. Some insiders suspected that one of the reasons why the SBIRS early warning satellite program was delayed for so long was that its funding was diverted into the F-22 and F-35 fighter jet programs. There is no direct evidence of this, but there are indications that in the past this indeed happened.

Fungibility at the big defense contractors is even harder to prove. The fact is that firms like Boeing and Lockheed have an incentive to keep the top Air Force leadership happy, and this is not accomplished by giving military space programs priority over fighters, bombers, and other air warfare assets. The big firms have never really been able to integrate space activities into their work. They are conglomerates, managed by business professionals whose main goal is to keep their operations and their jobs; profit is secondary, and everything else is tertiary or less. Building rockets, satellites, and space ground systems is hard and, even worse from senior management's perspective, it's risky. All too often the big firms choose not to bid on military space contracts knowing that if they win, they could spend billions of their own funds and still end up losing money.

It is ridiculous to expect the space industry—or any other industry—to do charity work for the Defense Department, yet all too often that is what Congress and the military demand. No wonder that firms reach for every excuse they can find to squeeze every possible buck out of their contracts and out of the immense and, until recently, ever-expanding pile of procurement regulations. The unfortunate habit of firms filing "protests" after major contract awards not only bites into everyone's time, but it eats away at the winners' profit margins. No wonder that one of Norman Augustine's most infamous laws held that "Bulls don't win bullfights, people do. People don't win people fights, lawyers do."

Today, most of those leaders and observers who've come out against the Space Force idea admit that eventually we are going to have to establish a separate service to manage our nation's military spacepower. In 2019, the Defense Department came up with a proposal that was practical and constitutional. Congress reshaped the Trump administration's plan to suit its own perceived interests, and the USSF was created. The timing was lucky, and the hysterical political atmosphere actually served to neutralize the usual anti-military forces; the people involved were, at the time, so wrapped up in the Russia Hoax kerfuffle that they failed to fight hard against a proposal that would inevitably strengthen America and weaken her foes. Space warfare is not going away; in truth, it is just getting started.

Working out what exactly space warfare is will not be easy. For example, is a cyberattack on a satellite control ground station an example of space war? But working out a practical definition is an essential first step. We tend to think of military airpower as largely consisting of putting bombs on targets and shooting down enemy airplanes, but of course it includes everything from troop and materiel transportation to air mail, casualty evacuation, and humanitarian operations. Once we understand the nature of space warfare, *then* we can work out a theory of spacepower—or decide which of the existing theories most closely matches reality. America's existing experience of space warfare is in some ways quite extensive, but it is limited to non-kinetic conflict.

Russia, China, Iran, and North Korea (and perhaps others) have been conducting unrelenting cyberattacks on US space assets with the goal of disrupting ongoing operations and preparing to shatter our systems in the event of a major crisis or war. So far, they have used cyberweapons, focused jamming, and occasionally low-powered lasers. The US government has chosen not to regard such actions as acts of war, but without question they come pretty close. The situation is "short of war" but often looks like a prelude to actual conflict. China's Strategic Support Force, which includes space, is an example of the way some nations are organizing their military services to fight wars and semi-wars in the Information Age.

This kind of political conflict, or semi-warfare, reminds one of the covert anti-commerce campaigns carried out by Elizabethan privateers against the Spanish Empire (of course, these days they rarely bring boatloads of treasure back to their supporters) or the efforts of the Confederacy to interfere with the Union's economy. As in the sixteenth, seventeenth, and eighteenth centuries, it's hard to tell the difference between interstate warfare and simple piracy. Just as Queen Elizabeth I could, with her tongue firmly in cheek, deny that she had anything to do with the activities of Francis Drake or the other English sea dogs, so Putin, Xi, and the rest issue laughable official statements claiming they are innocent of waging cyberwar against the rest of the world.

Our age is one in which political and economic warfare of various sorts are endemic. It should not surprise anyone that China, Russia, and America's other foes conduct unrelenting attacks on all forms of US national power, and this most emphatically includes spacepower. This is why the Space Force will be fighting a space war right from the start. It will, in fact, take over the ongoing warfighting missions of the Air Force and the rest of the military space forces. Training and equipping people and units who can conduct this kind of warfare, and doing so effectively in a classified environment, is going to be one of the first challenges the Space Force will have to face.

This also raises the question of how to organize to fight the cyberwar. Do we, in fact, need a Cyber Corps? Has the Air Force done an adequate job in cyberspace? The answers to these kinds of questions are impossible for someone without a very high-level security clearance to determine. The Snowden leaks showed that the National Security Agency (NSA) was doing a fair job of collecting massive amounts of data; there is no indication if the agency is or isn't using it effectively. Other leaks showed that the US has indeed been working on offensive cyber systems, but again, with no evidence one way or the other about how effective these cyberwarfare products really are. What we do know is that war in cyberspace, just like war in outer space, is an ongoing reality.

One reason why the USSF is essential is the inherent weakness in the "Joint" system of command, strategy, operations, and, above all, budgeting. Previously, military (and intelligence) space operations were spread out among more than a dozen different entities. While the Air Force may have had the lead role, it was just first among equals. The Navy and Marines, the Army, and civilian agencies such as the State Department and its various elements, the Commerce and Transportation departments, Homeland Security, and especially NASA, all had a say in military space policy. By itself the Air Force was too weak and too distracted to make a concentrated and effective policy push for American spacepower. Even with the goodwill and support of 99 percent of the officers, civil servants, and political appointees involved, it only took a single obstreperous bureaucrat, sometimes in a minor position, to sabotage a program or policy supported by the other 99 percent of people involved.

Over the years, the struggle over spacepower has engaged factions inside the government who are ready to go all out to protect their political investment in certain space policies, notably the idea that keeping space a demilitarized "sanctuary" is a supreme US national interest. These factions have successfully ensured that China and Russia have been able to go ahead with the development and perfection of kinetic space weapons while the US has done nothing. One reported example of this is the story that a single Air Force officer held up funding for the now-legendary DC-X program because he or she saw it as a threat. The program began during the George H. W. Bush administration paved the way—indirectly in the case of Elon Musk's SpaceX and more directly in the case of Jeff Bezos's Blue Origin—for the emergence of the reusable launch vehicle industry, which has revolutionized the cost of access to orbit. In 2021, this industry is a critical US spacepower asset. It is sheer luck that the program survived long enough to prove that the concept was viable.

One of the best examples of the "sanctuary" argument is the article by Michael Krepon, Theresa Hitchens, and Michael Katz-Hyman, "Preserving Freedom of Action in Space: Realizing the Potential and Limits of U.S. Spacepower." They make the usual case that since the US has the most to lose in any space war, it should refrain from doing anything to "weaponize" space, and they propose the usual weak set of international agreements and codes of conduct that would supposedly ensure this. What is most interesting about their proposal is that, recognizing that a purely arms-control approach is unacceptable to anyone with the slightest interest in US national security interests, they want the US to "hedge" its position.[6]

However, the low-level space war that is happening now may, at any moment, escalate into something closer to active conflict. The Space Force leadership, especially those involved at the operational planning level, will have to be prepared to deal with this. Defensive and offensive options must be available, and planners must be forced to take into account foreseeable enemy and allied responses to any US actions. Any US active military moves in space will, as of now, be unilateral, and those of our allies who've committed themselves to multilateralism will have to be reassured.

For example, what if one of our large and very expensive spy satellites were to be destroyed in low Earth orbit (LEO) by a space mine—which we suspect was planted by China? Could the US convince its allies to go along with a program of retaliatory measures that would inevitably harm their commercial relations with the Middle Kingdom? Even if Washington shared the proof with, say, Germany, do we really imagine that Berlin would go along with whatever the US found appropriate as a response?

Today, a president's options would be limited. Without a US anti-satellite weapon, we could perhaps respond by jamming Chinese satellites or by launching a cyberattack, but these actions would be seen by China and by the rest of the world as weak indeed. A limited cruise missile attack on the Chinese homeland would escalate the situation and perhaps lead to things getting out of control. The Chinese leadership might figure that it would be worth the risk to take out a large US satellite. Humiliating the Americans in space, a domain where most of the world had assumed US superiority, would be a major propaganda victory.

The USSF, if equipped with a variety of anti-satellite weapons both kinetic and non-kinetic, would give the president a set of proportionate options. No need to under- or overreact; the USSF could order the destruction of one or more important Chinese spacecraft. Having this option would, in and of itself, be a deterrent. However, a sophisticated and well-funded lobby has been operating in Washington for decades that has successfully prevented the US government from taking a realistic stance on the issue of space warfare. Thanks to the pressure from Trump and other supporters of the Space Force idea, the Air Force is changing its attitude toward space warfare. One example of this is the April 9, 2019, speech by USAF chief of staff Dave Goldfein when he said, "It's not enough to step into the ring and just bob and weave, block and parry, and absorb punches. At some point we've got to hit back." This is a blunt and seemingly unequivocal repudiation of the left's attitude toward US space power. However, without a Space Force to sustain and implement this new doctrine, America could soon find itself just as vulnerable the next time a liberal sympathetic to the arms controllers takes over the White House.

One reason, among others, for the success of the arms control lobby has been that the Air Force and its supporters in Congress and elsewhere have not been able to focus on the needs of the space warfighting community. Instead, Air Force leaders have either ignored the issue or adopted various delusional doctrines such as the idea that space is a sanctuary, or that if America builds a resilient set of space assets, the US military will not have to prepare itself to conduct "active" warfare in space. From the Air Force point of view, fighting a political battle over "space weaponization" would be a distraction from the need to fight battles to rebuild the nuclear bomber force and the ICBM force, which are core USAF missions. Only an independent Space Force with leaders who have the guts to force Congress to pay attention and to fight the long and hard political battles on Capitol Hill will be needed to get funding for a real space warfighting capability.

In the short to medium term, say, from 2021–2035, the field of action is going to be the Earth-Moon system. For now, all the targets of interest are either in geosynchronous equatorial orbit (GEO, 22,000 miles up) or closer to Earth, but it is inevitable that the US and other governments will place satellites farther and farther out—partly to make them more difficult to detect and attack, and partly because there will be more civilian assets out there which need to be tracked, monitored, and protected. Space telescopes are hugely expensive assets which may or may not have military uses in emergencies. Some of these observatories will be placed at or near the Lagrange points; others may be in lunar orbit.

The US has often considered having "on-orbit spares" in order to be able to quickly replace national security spacecraft that may be destroyed or damaged. The logical place for these satellites is beyond GEO, and it would be foolish to ignore the advantages of having a number of anti-satellite systems or so-called "space mines" hiding out there, ready to be used if needed. The Space Force will have to learn to track objects throughout the Earth-Moon system, just as the Air Force has learned to track objects in low Earth orbit. Achieving this goal may be more a matter of developing expertise than investing in new systems—that should be a question for Space Force strategists.

There is also the question of space mining, space industrialization, and space settlement. Both the US and China are seriously interested in building bases on the Moon, and Elon Musk's effort to colonize Mars is getting underway. Some of these ventures will be led by NASA; others will be led by private companies. The role of the Space Force in providing Coast Guard- or Army Corps of Engineers-type services to these projects will be one of the ways the USSF will directly support America's peacetime national interests. The sooner the Space Force is established, the sooner we will see America's commercial and civil space settlements flourish throughout the solar system. US economic interests on the Moon, Mars, and elsewhere will thrive under the protection and, one hopes, the loose regulation of the Space Force's Space Guard, just as our fisheries are, more or less, protected by the Coast Guard.

Naturally, the Space Force will not interfere with the legitimate military, commercial, and scientific interests of other nations—including those of our potential adversaries—but it will keep an eye on them. There is no way that an outsider can tell just how good a job the US intelligence community is doing at surveilling and understanding what the rest of the world is up to in outer space. The record is not too promising. The Space Force intelligence service may, by its very nature, do a better job. However, it is worth pointing out that the drive for more and better Space Situational Awareness (SSA) that has been supported by various left-wing organizations is not aimed at giving us a truly better idea of what nations like China and Russia are up to, but instead is aimed at preventing the US from developing its own "space weapons." The argument is that if we have SSA, we can order our spacecraft to avoid enemy anti-satellite weapons (ASATs). This means that an enemy could render our satellites useless simply by forcing them to expend their fuel in an attempt to avoid ASATs, both real and simulated.

Perhaps using very large numbers (in the thousands) of small, low-cost satellites could make conventional ASATs useless, but we're years away from having the kind of capability needed to replace the advantages we get from our "exquisite" satellites with "swarms" of microsats. However, using a number of the small satellites with today's capability to supplement the Keyhole imagery satellites is an obvious, good idea. The National Geospatial-Intelligence Agency (NGA) is already making use of commercial imagery satellites, and these programs should continue either under the USSF aegis or, as things stand, under the intelligence community.

Whether a Space Force comes into existence or not, the services, the intelligence community, and other governmental organizations (including NASA and NOAA) will continue to develop and build their own space-based information gathering systems—partly for reasons of prestige and institutional empire building, and partly because the leaders of these organizations don't think that outsiders can be trusted to give their needs the priority they think they deserve. There is a school of thought, best represented by former defense secretary Robert McNamara, that believes such scattered programs are inherently wasteful and that centralization is the best way to manage things. Others think that having multiple public and private organizations conducting their own projects will give the nation a better chance of having at least one or two outstanding successes, and that having a number of programs helps avoid the "too big to fail" problem that often plagues US government technology development programs.

One technology which has been completely ignored by the Air Force (and the rest of the government) is the idea that very large, very lightweight structures could be placed in orbit to serve as either communications satellites or radar imaging satellites. The antennas on the current generation of all-weather, radar spy satellites (Topaz) are roughly fourteen meters wide, but Ivan Bekey, a legendary space engineer, proposed that we build space-based radio frequency aperture antennas as big as 1,000 meters across.[7] Radar imaging spacecraft equipped with this type of antenna could be placed much farther from Earth and still do the same job as current ones, or, if placed in the same kind of 1,000-kilometer orbits as are currently used, they could send images that are far more detailed than we get today. These antennas have the capacity to change shape; this not only means that they can be adapted to various roles and missions but could also survive some types of attack. Innovative ideas like this will, in all probability, get a more respectful hearing from the Space Force than they ever could from the Air Force.

While Russia has not been making much progress in developing reusable launchers—and its new Angara series does not look too promising as a commercial proposition—China shows every sign that its leaders recognize the need to develop RLVs and have turned to the entrepreneurial sector to fill this need. Of course, in China, no major capitalistic venture can take place without the approval of the Communist Party, therefore the new "private" launch ventures can expect to be supported by the state. The US should therefore expect that sometime in the next decade China will have the means to compete with SpaceX and Blue Origin in the international commercial space launch market. In military terms, this means that China will have a low-cost way of getting its assets into orbit. When this happens, our nation had better be prepared; only a well-established and mature US Space Force—with leaders who understand the technology, the environment, and the special kinds of tactics and strategies that military space operations require—will be able to protect America's vital interests.

However, the main reason why we need the USSF is that today "space" is, for America, more of a vulnerability than an asset. The US military, not to mention the US economy, depends on space systems for everyday operations; farming, banking, and air travel are only a few of the areas where space assets play an essential role. In a crisis or an all-out shooting war, space systems will have a major role in determining the outcome. Only a Space Force—unambiguously dedicated to US space superiority—will be able to change this vulnerability into a war-winning asset, thus keeping the peace.

In spite of the objections (some of them valid) to the numerous small wars, the relative peace brought about by the US military superiority that has existed since 1945 is worth keeping. The proportion of the human race that can be expected to die violently in war is smaller now than at any point in recorded history. Once upon a time, the Romans would *ubi solitudinem faciunt pacem appellant* (they make a desert and call it peace). And that had been the norm for most of human history. Military superiority for one power meant misery and subordination for others. For all its many flaws, the *Pax Americana* has permitted an unparalleled flourishing of human civilization. With any luck, US military space superiority provided by the USSF will give our civilization another century or more of relatively peaceful development.

Some historians claim that because European colonialism has disappeared, the "Vasco da Gama Era" is over. Indeed, the political results of the fifteenth-century Portuguese explorers' voyages have been liquidated, though one may argue that the cultural and economic impacts echo on. Today, as we see SpaceX building ever more ambitious and efficient reusable rockets, historians in the future may write that the early twenty-first century was the "Elon Musk Era." If the new civilization that emerges is to reflect American interests and values, we need a Space Force that will support the US part of this development. In thirty or forty years there will be numerous American and non-American settlements, mining facilities, and industrial establishments on the Moon, on Mars, and elsewhere in the solar system; protecting them and keeping them safe will, in the end, be the primary mission of the Space Force—and ultimately the Space Guard (in effect, a Space Coast Guard). For the moment, the Space Force may be focused on supporting the terrestrial warfighters, but this focus should not prevent today's space service leaders from thinking about what kind of a future they want for their successors and for the nation.

Humanity is taking the first, tentative steps toward becoming a multi-planet species; the Space Force will only have a very small direct role in this outward push. But keeping the peace here on Earth will help create the essential foundation needed for the peoples of this planet to thrive, here and elsewhere in the solar system.

CHAPTER 2

A HISTORY OF US MILITARY SPACE

PART 1: A NEW ERA

There is a small bit of my family history connected with the beginning of the Space Age. On April 30, 1945, on the West Bank of the Elbe River, according to the history of the US 29th Infantry Division, "The three precisely dressed German officers (an SS lieutenant colonel, a Major Division G-3, and a lieutenant, were a bit surprised to see the field jacketed Captain Dinerman (my father) approach them. 'What rank do your battalion commanders have?' they wanted to know. However, it was made patently clear to them that they could either talk to the captain or recross the river with their mission unaccomplished. Thereupon the ranking German officer called Captain Dinerman aside and very carefully showed him a paper—an authorization from his division commander to surrender the entire division numbering approximately ten thousand troops. In half-whispered English he said that this was the V-2 Rocket Division, and that the Germans wanted to preserve the secrets of rocket propulsion for Western Civilization. Preservation of German rocket secrets was, of course not the responsibility of the 29th Division but accepting the enemy's surrender fell within the scope of its wartime duties."[8]

The V-2 began the Space Age. Theoreticians such as Konstantin Tsiolkovsky and Hermann Oberth showed the way, and Robert

Goddard's experiments pushed forward the engineering, but it was the V-2 that proved that large rockets could actually be built using mid-twentieth-century technology. After that, in the US and the USSR, it was just a matter of providing the resources and giving the technicians (aka rocket scientists) the time and support to build larger and better rockets.

In his memoir, Boris Chertok, who later became one of chief Soviet rocket designer Sergei Korolev's key assistants, wrote, "Before 1945 neither we, the Americans, nor the Brits had been able to develop liquid-propellent rocket engines with a thrust greater than 1.5 metric tons.... By that time, however the Germans had successfully developed and mastered a liquid-propellent rocket engine with a thrust of up to 27 metric tons—more than eighteen times greater!"[9]

From a historical point of view, the definitive works on the V-2 were written by Michael J. Neufeld: *The Rocket and the Reich*, published in 1995,[10] and his biography of Wernher von Braun, published in 2007.[11] The impact of the V-2 as a weapon is debatable. Neufeld points out that "more people died producing it than died from being hit by it."[12] According to Neufeld, some 3,200 V-2s were fired at British and Continental (mostly Belgium) targets. From a purely economic viewpoint, the V-2, like many other military projects, turned out to be a waste of money, but as a political weapon it was not without effect.

The V-2 helped inspire the Germans, both soldiers and civilians, to fight on. It showed there was hope that the Nazis did in fact have available some "Wonder Weapons." It is impossible to know how much this delusory hope contributed to making Germany's last stand such a bloody one, but it is likely that the impact was very real, if not measurable.

On the other hand, the impact on the morale of war-weary Londoners was also very real. British leaders repeatedly worried about the effects of the V weapons on the spirits of Londoners, who'd endured the Blitz and long years of deprivation and stress. However, it is worth noting that prime minister Winston Churchill never pressured Allied commander Dwight Eisenhower to adjust his strategy to occupy the V-2 launch sites on the coast of Holland.

At the end of the war, the US not only got the V-2 Division, which turned out to control few "secrets," but also the von Braun team, and the Mittlebau-Dora concentration camp was occupied and mostly cleaned out by the Americans before it was handed over to the Soviets. The hardware from Dora and the people from Peenemünde, where the V-2 was developed, were shipped to White Sands, New Mexico, where the US Army put them to work building and testing the rockets.

The beginning of the missile age gave rise to two questions: What kind of warheads are we going to put on these things? When we do build a missile that can put things in orbit, what kind of things should we send up? The answer to the first question was simple—whatever can do damage to our enemies. High explosives and nuclear weapons were the preferred solution, though some nations chose to put chemical and even biological weapons on top of ballistic missiles.

The answer to the second question was not self-evident. The head of the US Army Air Forces, General Hap Arnold, was interested in the future. In 1944 he said, "We have won this war and I am no longer interested in it.… Only one thing should concern us. What is the future of air power and aerial warfare? What is the bearing of new inventions such as jet propulsion, rockets, radar, and the other electronic devices?"[13] The next year he began Project RAND, which evolved into today's RAND Corporation. One of its earliest studies included a reference to the possibility of using Earth-orbiting satellites for reconnaissance purposes. In England, meanwhile, Arthur C. Clarke had published his revolutionary idea for space-based communication stations (see chapter 4).

In any case, defense budgets in the years from 1945 to 1950 were not very generous. Writing in 1958, General Bernard Schriever explained that the "Air Force ballistic missile development program was kept at a relatively low level until 1950…"[14] The first Soviet nuclear test, followed by the outbreak of the Korean War, changed things. The US military began to take seriously the capabilities inherent in this technology.

The 1952 election led to a dramatic change in priorities inside the Pentagon and the intelligence community. Eisenhower, who had benefited from an excellent flow of intelligence—mostly from the British and their code breakers—as supreme allied commander of NATO and as US president, found his sources of information about the communist world were almost non-existent. Making decisions based on intuition, wild guesses, or wishful thinking was not his style. He needed reliable sources, and in the absence of cryptographic intelligence, only overhead photography could provide him with the information he needed for sound decision-making.

Most importantly, he understood that as long as America lacked reliable intelligence about what was going on inside what was then called the Sino-Soviet bloc, we were vulnerable to a surprise attack—a second Pearl Harbor. In early 1954, Ike decided to establish something called the Technological Capabilities Panel (TCP). Its writ was to find ways of "increasing our capacity to get more positive intelligence about the enemy's intentions and capabilities and thus to obtain, before it is launched, adequate foreknowledge of a planned surprise attack."[15]

The TCP led to the development of the U-2 and the SR-71 spy planes, but panel members recognized that these were just stopgap measures and recommended the US develop spy satellites. This began under the auspices of the WS-117L program. Later, the Corona program was spun off from WS-117L and eventually became the first successful satellite to return film to Earth, providing the US—for the first time—with a comprehensive overhead view of the USSR.

Eisenhower ordered that the Corona program be accomplished under the control of the Central Intelligence Agency. He was wary of exclusively military space programs. He also insisted that the program be kept as secret as possible. It took more than a few failures before the first successful mission was carried out in August 1960—during the election and in the middle of the controversy over the "Missile Gap."

Corona proved the Soviets lacked the large ICBM force the intelligence community feared and that the Democrats used to hammer the GOP.

A good example of the way the president tried to keep space technology in civilian hands was the way he founded NASA as a specifically civilian agency. The fact that NASA was largely staffed by either military men or by civilians associated with the aerospace industry made little difference. NASA was and is a civilian organization that is as open and transparent as any US government entity can be.

What is perhaps most significant about Eisenhower and the Corona project is the speculation that he deliberately delayed America's first space launch and allowed Soviet Russia, in October 1957, to go first. While Sputnik shocked America and the world, it also allowed the US to establish the "Freedom of Space" principal; by not complaining about Sputnik when it passed over US territory, the administration opened the way for US satellites to pass over the USSR in all legality.

It's true that the administration did restrict the von Braun team's ambitions, but was this more due to the president's resentment of the German for his wartime role (and possibly also his role promoting an ambitious and very expensive space exploration plan at a time when Ike was trying to balance the budget)? Or because he was playing 3D chess? The evidence is mixed, but personally, I like to believe it was an example of "the Hidden-Hand Presidency."

On the other hand, there is lots of evidence that America's failure to be the first nation to launch a satellite was due more to the all-too-common muddles that happen when politics, bureaucracy, and technology intersect. After the usual flops and blowups, von Braun successfully launched a Jupiter-C intermediate-range ballistic missile (IRBM) in May 1957. His team quietly went ahead and prepared to use a version of that rocket to put a small satellite into orbit, but when this became public, John Bruce Medaris, von Braun's boss, "issued decrees stopping all work on satellite configurations of the Jupiter-C and even upbraided von Braun for having let things go beyond his orders."[16]

In the mid- to late 1950s, both the US and the USSR were engaged in comprehensive missile development programs, mostly geared toward the European theater. Their motivations, however, were very different. The US was trying hard to substitute nuclear firepower for the fully manned and equipped divisions that NATO lacked. Fighter bombers with atomic gravity bombs were regarded as a stopgap measure; they were vulnerable to attack on the ground and in the air. Medium- and intermediate-range missiles that could be moved around and hidden were seen as a much more credible deterrent. Soviet Russia, on the other hand, lacked an advanced aerospace industry; its aircraft, while good, solid fighting machines, lacked the refinements of their American and Western counterparts. In Korea, the US F-86 Sabers had defeated the MiG-15s, and the men in the Kremlin had no reason to believe that if war came to Europe, their air force would do better. Rockets of all types with a wide variety of warheads would have to do the jobs that their inferior airplanes could not do.

America deployed many of these missiles in places like Turkey, Italy, and the UK. They were withdrawn in the early 1960s due to local governments' fears and as part of the secret deal that ended the Cuban Missile Crisis. Later, in the late 1970s, European governments asked president Jimmy Carter to deploy a new generation of US missiles to respond to increased Soviet deployments.

One bizarre point is that the US did have a small but significant lead over the Soviets in solid rocket fuel technology. This was, in part, due to the work of John Whiteside Parsons, a self-taught explosives expert and part-time Satanist and sex-cult member, whose work with the Caltech-centered "Suicide Squad" helped lead to the creation of both the Jet Propulsion Laboratory (JPL) and the Aerojet corporation.[17]

America's lead in this technology led directly to both the submarine-launched Polaris missile and to the silo-based Minuteman ICBM. Both weapons and their successors gave the US an advantage in reliability that the USSR was never really able to overcome. These US advantages were never discussed outside a few specialized publications.

The Missile Gap argument was used ruthlessly by both the Democrats and the media (in those days, they were slightly less joined at the hip than they are now) against the GOP and Richard Nixon.

In December 1957, less than two months after Sputnik, the US successfully tested its first ICBM, the liquid-fueled Atlas. It was only in October 1960 that the missile was actually deployed, and by 1962, more than 120 were operational—far more than the Soviets had available. By the end of the 1960s, most US liquid-fueled ICBMs had been retired. It turned out that they were far better as space launch vehicles than as weapon delivery systems.

In any case, John F. Kennedy won the 1960 election, and after a rough start to his presidency (including the Bay of Pigs disaster and a humiliating summit with Soviet premier Nikita Khrushchev in Vienna, not to mention Yuri Gagarin's first manned space flight), Kennedy ordered NASA to "land a man on the Moon and return him safely to the Earth." From a military point of view the new president chose a Republican technocrat, Robert McNamara, as his secretary of defense.

All SecDefs tend to be control freaks, but McNamara was exceptional greedy for raw bureaucratic power. He scrapped the doctrine of massive retaliation, which had the virtues of keeping the size of the defense budget more or less under control and, more important, kept all nuclear decisions exclusively in the hands of the president.

The new doctrine of flexible response, which forced the president to rely on committees of experts, needed better, more secure, and more survivable communications, and this gave extra impetus to the military's communications satellite programs. The Navy in particular needed these systems, as well as space-enabled navigation for its Polaris missile submarines.

One piece of advice that the outgoing president gave to his successor was never to underestimate the importance of the Polaris missiles. They were the most survivable nuclear systems available and could be kept in reserve to threaten retaliation, even after a Soviet first strike.

This was also a (brief) period when the armed services were allowed to think imaginatively about space warfare. Project BAMBI (Ballistic Missile Boost Intercept) for example, proposed to base an interceptor in low orbit from which it could hit enemy missiles in their boost phase, an idea that refuses to die no matter how often the civil/military establishment tries to kill it.

In response to a Soviet anti-satellite test, the US tested and deployed its own nuclear-tipped ASAT weapons, based first on Kwajalein Atoll in the central Pacific. Later, a different version was installed on Johnston Island farther east. These were both stopgap measures, intended to show the Soviets and the American public that the US defense establishment was doing *something*.[18] Unfortunately, the US government neglected realistic kinetic ASAT weapons for decades in the stubbornly held but utterly delusional belief that US restraint in this area would be reciprocated. It turns out our foes believed we were developing and deploying secret ASAT weapons or that we were being naive.

The idea that space could be kept as a "sanctuary" was an article of faith held by arms control advocates and senior military people alike until it recently became unmistakably apparent that Russia and China had built a comprehensive array of space war weapons. The reasons for this belief were, in a perverted way, perfectly logical. Since the US had more on-orbit assets, it had more to lose if a space war broke out, so by showing restraint, Washington would discourage Moscow, and later Beijing, from building such weapons. This would avoid an "arms race in space," which would be expensive and would benefit no one.

It turned out the men in the Kremlin thought that having the capacity to destroy America's space assets would be a useful thing. So they pursued this type of weapons development with few restraints. In any case, the limited US ASAT system based on Johnston Island remained operational (just barely) until it was dismantled in 1975.

The US came to depend on spy satellites for most of its information on Soviet strategic forces. One important role was that they allowed CIA analysts to have a reliable technical source they could use to see if the few human sources they had inside the USSR were telling the truth. This development took a long time to perfect, since the imagery from the Corona satellites was, at first, not very abundant and was so highly classified that it was not easy to use. On top of this, the photo interpreters had a pretty steep learning curve. It took years for the US intelligence community to build the skills needed to get the most out of the pictures the satellites were taking. These skills were centered in the new secret intelligence agency the National Reconnaissance Office (NRO).

Fortunately, by 1964 the vast majority of Corona missions were successful. This allowed the image interpreters to build up libraries of pictures that provided histories of what various sites looked like over time. However, the spy satellites had serious limits, and the Soviets knew all about them. Khrushchev is quoted as telling French president Charles de Gaulle, "As for Sputniks, the US has put one up that is photographing our country. We did not protest; let them take as many pictures as they want."[19]

This meant that the Soviets knew our satellites were flying over their empire and they could track them. Americans learned, as other nations had learned before, that Russia is very good at hiding things— especially military things—it does not want the rest of the world to know about. Just how good or how bad America's image analysts were is still unknown, but it should be obvious that the leaders in the Kremlin did not make it easy for them.

Spy satellite missions were almost all devoted to gathering what can loosely be called "strategic intelligence." As far as we know, these missions contributed little or nothing to America's war in Vietnam. Photo intelligence during that conflict came from modified tactical aircraft or from planes like the SR-71 Blackbird.

In the 1970s, according to the story, George H. W. Bush ordered the NRO to begin work on radar imaging satellites colloquially known by the code word "Lacrosse." Their ability to see through clouds makes them particularly valuable; they fly in LEO and they have often been spotted and photographed by amateur space trackers. Like the super-capable optical Keyhole satellites, there are at most only two or three of these expensive and hard-to-operate items in service at the same time.

In any case, imagery satellites may have been the first US intelligence-gathering systems in space, but they were soon followed by ones involved in what loosely can be termed "electronic intelligence" gathering. War in the electromagnetic spectrum has been going on since the Japanese navy discovered it could jam Russian radio signals during the early days of the Russo-Japanese War in 1905. In World War II, collecting transmissions and breaking codes was critical to the allied victory.

Meanwhile, the use of radar led inevitably to the development of various radar countermeasures. The ability to jam enemy radars depended on knowing what frequencies those radars were using. Gathering data on radar frequencies was sometimes a difficult and dangerous job. Specialized Royal Air Force (RAF) aircraft equipped with detection equipment would fly missions over Germany to collect and record the information.

With the coming of the Cold War, gathering data on Soviet air defense radars became infinitely more difficult. A few electronic intelligence-gathering missions were flown by the USAF and the RAF, but these were provocative and so often led to the loss of the aircrew involved that the Eisenhower administration eventually stopped them. However, the need for what is called "electronic order of battle" intelligence did not go away.

The Air Force and the intelligence community, including the super-secret National Security Agency, naturally looked to space as a location from which they could overfly the Soviet bloc, especially deep inside Eurasia, where the enemy could test new radars and other electronic systems far away from the prying ears of the West.

At first, small so-called "ferret" satellites were launched as well as a variety of electronic intelligence secondary payloads, but in the mid- to late 1960s the US began to develop and build larger specialized eavesdropping spacecraft. Early examples included the Rhyolite and Aquacade satellites, followed by the Mentors and, now, by the apparently very large Orions, which use antennas as large as 100 meters across.

These highly classified satellites are all controlled by the intelligence community; the Air Force's role is limited to providing launch services and occasionally hosting ground facilities at some Air Force bases. This nice, neat arrangement has come under stress as it becomes obvious that space is a warfighting domain, but the intelligence people and America's arms control diplomats have resisted the implications of this reality.

One vulnerability that the arms control advocates rarely mention is that these systems can be exposed, either by espionage (the "Falcon and the Snowman" case in the mid-1970s exposed the Rhyolite spacecraft's capabilities) or by leaks to the press. Many of the worst of these happened in 1975–1976 during the hysteria over so-called "CIA misconduct." Then-secretary of state Henry Kissinger later made the point, "The most pointless and damaging disclosures concerned technical intelligence.... Various communications intercepts, such as our ability to listen to some of the Soviet leaders' telephone conversations, were also pointlessly disclosed. These reevaluations involved intelligence successes, not failures, and carried not the slightest implication of abuse of authority."[20]

A few short years after this damage was done, the Carter administration tried to sell its SALT II arms control deal to the US Senate on the basis that it could be verified using "national technical means." For some reason, the US government still believed it was in its interest to deny the existence of the NSA, but did, at last, formally acknowledge the reality of America's "photoreconnaissance satellites."

At the time, the battles over arms control deals were the main theater—in every sense of the word—of Cold War political operations in Washington. Decisions on whether to build nuclear missiles, bombers, and their whole supporting infrastructure were shaped by the need to kowtow to the Gods of Nuclear Disarmament and rarely to the actual military needs of the country. This situation locked America into a losing position vis-à-vis the Soviets. They built up their forces with little regard for what the US did or for their obligations under the various treaties they'd signed, such as the Anti-Ballistic Missile Treaty or the Biological Weapons Convention. It was always easier for the USSR to develop and build newer and more powerful missiles and to deploy them than it was for us. Soviet leader Leonid Brezhnev never had to sign an environmental impact statement, let alone beg members of the Appropriations Committee for funds to build new weapons.

This situation changed forever in March 1983 when Ronald Reagan gave his famous "Star Wars" speech. It is sometimes forgotten that much of the Oval Office speech was devoted to a plea for the money to build the MX missile. The speech changed the very nature of the strategic competition with the communist empire; suddenly the main question was not who had the most and the scariest missiles, but who had the better technology.

Reagan had identified a deep philosophical problem with the mutually assured destruction (MAD) strategy. It did nothing to defend the American people. Indeed, between the 1950s and, above all, during the McNamara period, it might have been better to call America's military establishment the Department of Revenge rather than the Department of Defense. Missile defenses might have been hard to build with the technology available in the 1960s and '70s, but aside from the desultory Sprint/Spartan systems which were operational for only a few months, the Pentagon and its masters in the White House and Congress did almost nothing to develop technologies that would protect the US population.

Some historians trace the beginning of Reagan's attitude toward missile defense to a visit he made to the North American Defense Command's Cheyenne Mountain bunker in Colorado. At the time he was shown the capabilities of the early model Defense Support Program (DSP) satellites, which could detect even small missile launches from their position in geosynchronous orbit. The future president supposedly asked that if we had such great technology, why couldn't we stop the missiles rather than just detect them?

Others, who had a better grasp of the way American conservatives had been thinking about nuclear strategy, knew that even before the days of Barry Goldwater's 1964 presidential campaign, right-wing anti-communists had been pushing for a serious program of homeland defense to complement US offensive missile and bomber forces. Supporters of the GOP candidate like Phyllis Schlafly complained that the Kennedy-Johnson administration had canceled programs like the "Nike Zeus and Nike X missile killers."[21] The conservative wing of the GOP had never accepted the idea of MAD as the end point of US nuclear strategy.

Reagan went well beyond anything that conservatives had imagined. He set out an extremely idealistic goal of making nuclear weapons "impotent and obsolete." He must have known that in the short or medium term this was, as the British say, "not on." However, he did push the Defense Department into establishing the Strategic Defense Initiative Organization (SDIO), now called the Missile Defense Agency (MDA). This turned what looked to many like a crazy dream into a bureaucratically established institution that would fight for its share of the budget and for its own survival. Since 1983, US missile defense programs have made progress, and, if properly funded in future years, will produce systems that will degrade the effectiveness of even the most advanced Chinese and Russian offensive missiles. It may take a century or longer for Reagan's vision to become a reality, but as long as the MDA exists, the US will continue to take small steps in that direction.

In any case, what made Reagan's SDIO such a threat to both the USSR and the US arms control establishment is that there already was a realistic proposal to build a space-based missile defense which, if pursued, would neutralize a big chunk of the Soviet land-based heavy missile force as it then existed. This was the High Frontier plan put together by retired Army General Daniel Graham and his team.

High Frontier should not be confused with the High Frontier plan put together by Gerard O'Neill of Princeton University espousing space colonization and space solar power. O'Neill's ideas have influenced many people over the years, not only people like Elon Musk and Jeff Bezos, but also quite a few military people who support both space-based missile defense *and* space solar power.

Dan Graham's ideas were boiled down to the concept of "smart rocks." In his 1983 book, Graham described the core of his proposal: "A representative GMD [Ground-based Midcourse Defense] system includes a large number of satellites, or 'trucks,' distributed in circular orbits at an altitude of approximately 300 nautical miles. The example here uses 432 trucks, all in orbits inclined 65 degrees with the equator."[22] Each truck would have forty or so heat-seeking interceptors.

It is important to observe that in the 1982 Falklands War, the Royal Navy and the Royal Air Force had proved the effectiveness of the American Sidewinder AIM-9L. Heat-seeking missile technology sensitive enough to destroy enemy planes in the rough, unpredictable skies of the South Atlantic should logically be sensitive enough to detect the far greater heat put out by a heavy ICBM thrusting through the Earth's upper atmosphere. The Soviets obviously knew this, but Western leftists stubbornly refused to accept its reality.

What made matters worse from Moscow's point of view was the growing gap between the Soviets' computer technology and that of the West, particularly American microchips. In 1983, they were struggling to match the IBM mainframe systems of the early '70s, while in the US, personal computers made by Apple and IBM were proliferating into almost every middle-class home. Socialist science seemed unable to match the products of consumer capitalism, and the politburo was acutely aware of the military implications of this reality.

The men in the Kremlin might have been confident that with enough time, effort, and money they could develop countermeasures to the proposed US system, but since they were already under massive economic pressure to pay for their already-large military force and their global imperial socialist bloc, they saw SDIO as a major threat. They were already engaged in a gigantic political "peace" effort to prevent the US deployment of new missiles to NATO and to build up the nuclear freeze movement in the American homeland; adding a theme to defeat "Star Wars" and to stop the "weaponization of space" seemed an obvious and low-cost countermove.

Like many of the political battles that were fought in the last stages of the Cold War, the Soviets had a huge cultural advantage. It was child's play to insert anti-SDIO messages into movies like *Robocop* or *Mindwalk*, but as British political commentator John O'Sullivan wrote of the "peace" strategy: "At its height it could bring millions into the streets overnight. It reshaped the political culture of the left and thus of countries where the left was culturally dominant. It was so successful that it was imitated wholesale by the American left in the 'nuclear freeze' movement. It could do anything, in fact, except win elections."[23]

What most exasperated the opposition to SDIO was Reagan's optimism. Indeed, if there was one thing the president excelled at, it was communicating a message that America's greatest days were ahead of her. That America could accomplish seemingly impossible tasks and that the old "can-do" spirit was still alive and well. In the aftermath of the Vietnam defeat, not to mention Watergate and the oil crises of the 1970s, this seemed to many on the left and a few on the right to be a form of blasphemy against "realism."

Many conservatives are still bitter about the way the Reaganites were purged from the government after the 1988 election. On the issue of missile defense, George H. W. Bush was completely faithful to Reagan's vision. He put Reagan's ambassador to the space and missile defense talks in Geneva, Hank Cooper, in charge of the SDIO. Cooper took the promising "smart rocks" idea and turned it into the Brilliant Pebbles program, which passed a number of critical engineering reviews in 1991 and '92.

Cooper also played godfather to the DC-X experiment's reusable rocket project, which helped pave the way for Elon Musk's SpaceX launch systems and, more directly, set the stage for Jeff Bezos's Blue Origin launch vehicle development programs.

To sum things up, SDIO was the most significant change the Reagan administration inspired in the US military space posture, but it was not the only one.

PART 2: GROWING PAINS

When in 1982 the Reagan administration set up the US Space Command as a joint command based in Colorado, US military space seemed to be on a smooth path to future success in all areas. The Reagan administration broke with previous ones when it pursued the Miniature Homing Vehicle (MHV) ASAT program. This was a small, direct-ascent ASAT that took advantage of the F-15 fighter jet's amazing acceleration to boost itself so that it could target and destroy enemy satellites in low Earth orbit. It was tested once in September 1985, but the program was killed by Democrats in Congress in 1988. Since then, the US has chosen to avoid building an operational ASAT, though the technology is easily available and the Russians, Chinese, and, most recently, India have all been working on a wide variety of kinetic space weapons.

By the mid-1980s, decisions made ten years earlier were about to have some painful effects on the whole of America's push for space-power. The space shuttle had been Nixon's single major space program. It was supposed to replace existing expendable launchers with a manned, reusable, and relatively low-cost spaceplane that would open up a new era in access to LEO. The shuttle indeed had space for seven astronauts, but it was never low-cost and it could only be described as reusable by stretching the meaning of that word beyond recognition.

It took four years of political and bureaucratic wrangling for Nixon's people to come up with an answer to the question, "What comes after Apollo?" By the time the president made the decision to go ahead with the larger shuttle option in January 1972, it was an election year and, naturally, the Democrats were in no mood to give Nixon even a mild symbolic boost.

Nixon promised "an entirely new type of space transportation system designed to help transform the space frontier of the 1970s into familiar territory, easily accessible for human endeavor in the 1980s and '90s."[24] In spite of the rhetoric, the development budget for the shuttle was too small to begin with, and the program was underfunded throughout the design and development process. The technology was based on what was available at the time.

The Defense Department was not exactly enthusiastic, but NASA had earned a high degree of credibility thanks to the Apollo Moon landings, so the leaders at the Pentagon were not inclined to push back against a project that the US civil space agency had its heart set on. It has often been claimed that the cargo bay of the shuttle was specifically designed to hold the big spy satellites the NRO was working on. This is partly true, but NASA also wanted the large bay in order to launch the segments of a yet-to-be-designed space station and for large probes and observatories like the Hubble telescope.

Jimmy Carter nearly canceled the whole program but, according to legend, agreed to proceed only because he wanted to use it to deploy satellites he deemed essential to verify the SALT II deal. Once the program was underway, it was assumed that the shuttle would replace all of America's expendable launch vehicles. In effect, the US government was putting all its eggs in one space transportation basket. It turned out to be an expensive bet that did not pay off.

In its own way the shuttle was a technological marvel. Its easy flights symbolized America's revival under Reagan, even though his administration had had almost no role in the program. For those who followed the program in detail, it proved disappointing to say the least. While the space agency had promised dozens of flights a year, it never achieved more than nine—and that was in 1985, the year before the *Challenger* disaster of January 1986.

That disaster not only cost the lives of seven astronauts, but it grounded the spaceplanes for almost three years. The investigations castigated a NASA culture which had lost the habits of engineering excellence that had characterized its glory days in the 1960s when the nation devoted almost a full percentage point of its GDP to the space program.

In his speech memorializing the lost astronauts, Reagan made it clear where he stood: "Today, the frontier is space and the boundaries of human knowledge. Sometimes, when we reach for the stars, we fall short. But we must pick ourselves up again and press on despite the pain. Our nation is indeed fortunate that we can still draw on immense reservoirs of courage, character, and fortitude—that we are still blessed with heroes like those of the space shuttle *Challenger*." Reagan, like any president who instinctively "got" why space is so important, was drawn automatically to compare it with America's pioneers.

In the aftermath of the *Challenger* accident, as in the aftermath of the 2003 *Columbia* accident, it is remarkable just how few voices were heard demanding that the US give up the whole enterprise. It is hard to imagine under what circumstances the US would shut down all of its space exploration programs.

Compounding the disaster from the military's standpoint were a series of launch accidents which grounded the older expendable launch vehicles. In the expectation that the shuttle would completely replace the Titan, Atlas, and Delta rockets, neither NASA nor the DoD had invested much in trying to keep these programs viable. Now, suddenly, it appeared the US had lost the ability to deliver payloads into orbit. At a moment when the Cold War was entering its delicate and dangerous final phase, this could have spelled deep trouble for America and for its leaders.

The loss of spy satellites on two Titan 34-D launch failures in August 1985 and April 1986 proved to be the most serious setbacks to the US military space effort since the early 1960s. They forced the

Defense Department to completely and expensively rethink and rebuild its launch capability.

The lost spy satellites were urgently needed to help keep track of the Soviet SS-20 missiles which were the cause of what became known as the Euromissile Crisis of the early 1980s. The 1987 Intermediate-Range Nuclear Forces (INF) Treaty could have been derailed if the US Senate had decided that, thanks to these launch failures, the deal was unverifiable.

When the shuttle resumed operations, the first few flights were dedicated to either purely military cargoes or quasi-military satellites such as the Tracking and Data Relay Satellite (TDRS), which served both NASA and the DoD. In 1989, the first year of the George H. W. Bush administration, the shuttle was back in business and almost everyone thought that NASA had learned its lessons.

Bush 41 was more devoted to America's space programs than any president since Lyndon Johnson. Sadly, unlike LBJ, he had to contend with a Congress controlled by the opposition party. The story of how the Democrats thwarted his domestic and some of his international programs is a shining example of what Congress can do when it is led by skilled politicians determined, above all, to wreck a president's plans and desires.

In 1989, on the twentieth anniversary of the Moon landing, Bush announced that America would go "back to the Moon and on to Mars." To fulfill this goal, NASA undertook what was called the Space Exploration Initiative (SEI), with the initial goal of developing technology that would, eventually, carry out the president's promise. The Democrats in Congress took an almost sadistic joy in hunting down and defunding anything remotely connected with the SEI. This had minimal direct influence on military space, but it certainly did nothing to help the US space industry improve itself.

As the US recovered from the disasters of 1985 and '86, the intelligence-gathering satellite system was improved, both with better spacecraft and, just as important, better interpretation technology.

However, at this time, the superpowers were losing their monopoly on space imagery. Low-cost sensors and the proliferation of national space launch vehicles were changing the way America's allies treated space-derived intelligence.

When Saddam Hussein's Iraq invaded Kuwait in August 1990, there were already a few non-US (and non-Russian) remote-sensing satellites—most notably France's SPOT 1 and SPOT 2 in orbit. The US quickly cut deals with Paris to buy the image products, and these turned out to be useful additions to the information from America's own spacecraft.

The first Gulf War is sometimes called the first space war. This may be questionable, but the one new space technology that impressed the world had little to do with intelligence and everything to do with navigation. When the war began in February 1991, sixteen early model GPS satellites were in orbit, not enough to give full coverage, but enough to change the nature of the war.

The Iraqis were convinced that no army could navigate through the desert area along the Saudi border. The allied deception plan, which included making them think the US Marines were going to make an amphibious landing and that the heavy Anglo-American tank and mechanized forces were going to head straight into Kuwait, was successful. The wide swing though the desert worked because every unit (and in some cases almost every vehicle) was equipped with GPS. No commander relied on fallible map and compass readings. The coalition achieved total operational surprise.

GPS had proven its military worth. It was soon to prove itself an indispensable part of global civilization, so much so that other nations developed a bad case of what could be called "GPS envy."

In 1991, the Patriot air defense missiles were used, with mixed success, against the Iraqi Scuds. It would have been impossible for this system to have even limited success without the early warning Defense Support Program (DSP) heat-detecting satellites based in geosynchronous orbit. The sensors on these spacecraft were not designed to detect the relatively weak heat signature from an operational tactical missile like the modified Scuds. Over the years, however, the DSP heat detectors had been tweaked to be able to see and classify all sorts of targets—notably the afterburners on Soviet Tu-22M Backfire bombers. The DSPs earned their keep and showed what the US military space establishment could do if given the right kind of backing.

However, in spite of his victory in the Middle East and his successful management of the disintegration of the Soviet Empire, which lead to America's victory in the Cold War (or do we now have to call it Cold War I?), Bush was defeated in the 1992 election. His virtues as a man and as an international statesman were effectively turned against him.

The winner, Bill Clinton, didn't much care for space—civilian or military. In particular, he allowed his first secretary of defense, Les Aspin, to kill the space-based missile defense system known as Brilliant Pebbles. Given the Democrats' ideological commitment to the MAD policy, this was to be expected. What was not expected was the way that many Republicans didn't see the fight for missile defense as being over. When they unexpectedly won control of both the Senate and the House in 1994, the Republicans, as a unified party, began to push for some sort of national missile defense, and Clinton, reluctantly, went along. By the end of his second term, the Arkansas Democrat had put in place a limited ground-based missile defense development program that was eventually picked up and supported by his successor, George W. Bush.

The most successful space program begun by the Clinton administration was the Evolved Expendable Launch Vehicle (EELV) program, which resulted in the Delta IV and the Atlas V rockets. The plain-vanilla version of the Delta IV proved too expensive for most purposes, but the Delta IV Heavy was essential for lifting heavy payloads such as big spy satellites into orbit. The Atlas V, with its Russian RD-180 engine and Swiss-made payload shroud, was and still is an extremely useful launcher for both the Defense Department and for NASA. One of its greatest successes was to place the New Horizons Pluto mission on its path to the edge of the solar system.

Its Russian engine is subject to controversy. In the original deal, the Russians were supposed to provide the US with plans to enable us to build our own copies of the RD-180. The plans they provided proved inadequate; they left out the "secret sauce." Later, as relations with Moscow deteriorated, congressional demands to stop buying the engines became too loud to ignore.

Another military space milestone was the first use, during the 1999 Kosovo campaign, of the GPS-guided JDAM (Joint Direct Attack Munition) bombs. Before JDAM, precision guided weapons were simply too expensive to use on a large scale. These bombs with simple guidance systems made it easy to accurately destroy lots of targets with less, but not zero, collateral damage.

It's arguable that the Kosovo War was the first one to be won purely by air power, though the Kosovars who fought on the ground might object. The political circumstances of the war were such that the usual left-wing, anti-war movements were mostly absent. The Serbs had only the Russians and a few, not very respectable, right-wingers for allies. Even so, if the JDAM had not given the US and its NATO allies the ability to strike accurately with very few collateral casualties, the pressure to end the fighting while the Serbs remained entrenched in their positions might have overwhelmed the Clinton administration.

As an aside, some Serbian patriots are supposed to have stood on the Belgrade bridges and shouted up at the US warplanes, "Bite, Monica, Bite!" Their confidence in the accuracy of America's bombs and in the sense of humor of our intelligence services is, in a strange way, touching.

Clinton's VP, Al Gore, had a reputation as a technophile. His record, however, is mostly one of failure. The ill-fated Future Imagery Architecture program, aimed at developing a new generation of spy satellites, is a good example. Gore took public responsibility for the X-33 spaceplane program, supposedly a shuttle replacement with a revolutionary linear aerospike engine. The prototype never made it out of the hangar, let alone off the ground. As a follow-up to the path-breaking DC-X program, it was a disaster.

The younger Bush shared his father's enthusiasm for space, but after having watched his father's SEI be used to humiliate him, he moved cautiously in the civil space arena. In the military space field, Bush 43's first—and most memorable—secretary of defense, Don Rumsfeld, came into office committed to improving America's military space forces. At the time (summer 2001), one Air Force officer, a space specialist, was heard to say, "At last, we have someone in the E-Ring who loves us." (The E-Ring is the part of the Pentagon where the top leaders have their offices).

Before the election, Rumsfeld had served on a commission on national security space issues "examining how our patchwork of national security institutions dealt with space."[25] The new defense secretary understood technology and was familiar with the difficult, bureaucratic ways of the military. He was also arrogant and lacked charm, but you can't have everything. When he took over, the new buzzword was "transformation," which eventually joined other buzz-words such as "revolution in military affairs" and "network-centric warfare" in the discard pile of US military slogans.

It was Rumsfeld's misfortune that he found himself locked in a rivalry with the most skillful practitioner of bureaucratic politics in the history of the republic, with the possible exception of J. Edgar Hoover. When Colin Powell took over the State Department after eight years of Clinton appointees, the response of one experienced diplomat was, "Finally! Adult supervision!" Combining the heft of membership in the four-star club with the twentieth-century equivalent of a log-cabin-to-the-top biography and a near perfect ability to manipulate the media, Powell was one of Washington's most formidable power players.

On space issues, he was mostly content to leave things to the professionals and to the White House. The 2001 decision to pull out of the 1972 Anti-Ballistic Missile (ABM) Treaty had Rumsfeld's fingerprints all over it, but Powell managed to give the impression that the move had nothing to do with him. The huge improvement in US-India space relations, which culminated in strong NASA support for India's Moon mission, likewise, was more of a White House move than a State Department one.

Powell positioned himself as the guardian of the foreign policy establishment against the wild "neocon" idealists in the White House and the DoD. There is no direct evidence, but he and his team may have been responsible for killing off the effort to revive the Brilliant Pebbles space-based interceptor program, which was on the road to reincarnation in the weeks leading up to the 9/11 attack. Preventing "space weaponization" is one of those establishment policy positions that has endured long after it became obviously counterproductive.

Rumsfeld's policy victories all turned to ashes when the first stages of the Iraq occupation failed. Between 2003 and 2006, military space policy was on automatic pilot. This lack of attention probably doomed the promising new communications satellite program, the Transformational Satellite Communications System (T-SAT), and also allowed the early warning DSP replacement program known as SBIRS (Space-Based Infrared System) to go through a series of cost overruns and delays that nearly killed the whole program.

When Obama won the 2008 election, one of the things he did to reassure the DC establishment and the allies was to keep Bush's last defense secretary, Bob Gates, on the job. In spite of an amazingly nasty and dishonest confirmation process in 1992 when George H. W. Bush nominated him to be head of the CIA, Gates remained a fully paid-up establishmentarian. He kept his cool and he kept the DoD system as it was. On military space policy he made no significant changes; the US failed to answer China's 2007 anti-satellite weapons test with anything other than a minor demonstration against an out-of-control US satellite in LEO. And even then, it was reported that the software used to modify the Navy SM-3 missile for the shootdown was discarded.

In passing, one should note that the act of blowing up the errant satellite did not produce a large cloud of orbital junk. It was carefully done in such a way that the debris burned up in the atmosphere. This was partly due to the relatively low altitude at which the shootdown took place, but the skill involved cannot be ignored. Somebody did an excellent job of calculating just when and where to hit the spacecraft so as to minimize any adverse effects.

When Obama won the presidency, one of the campaign promises he tried to keep was the idea of banning space weapons. It soon became evident just how impractical this was, but his administration did try to hobble the US by introducing, with a great deal of help from the European Union, a "Code of Conduct for Outer Space Activities." Throughout the eight years Obama was in power, the US allowed China and Russia to continue their space weapons programs unhindered.

When the GOP won control of the House in 2010, a few Congressmen and women began to push for a US space-based interceptor program as a follow-on from the old Brilliant Pebbles program. The Obama White House, with assistance from left-wingers inside and outside the government, managed to stop the effort in its tracks. Indeed, the Obama years were ones when "transparency and confidence-building measures" were often deployed instead of hardware.

The administration lacked the votes to push through any significant treaty deals. The New START Treaty, which went into effect in 2011, made no real changes to the US strategic posture, or to the Russian one. However, Obama and his team tried to get around the Senate's prerogatives by making deals that, in theory, had the authority of international law, but because they were "deals" and not treaties, these agreements did not need to be ratified in accordance with the Constitution. The small kerfuffle over the Code of Conduct for Outer Space Activities and the much better-known controversy over the Iran deal did their parts to undermine the few shreds of trust that existed between Capitol Hill and the White House.

From the perspective of American national and not just military spacepower, Obama's most important decision was the cancellation of Bush's Constellation project. This NASA program, designed to go back to the Moon and eventually to Mars, depended on three initial elements: the Orion capsule, similar in shape if not in technology to the Apollo capsule; the Ares I launcher, based on a single solid-rocket booster from the shuttle and the Ares V; and a giant rocket using shuttle main engines and a pair of enlarged solid-rocket boosters. The program had been carefully put together to satisfy both political parties, in both houses of Congress, as well as the traditional aerospace industries.

When the whole program was canceled in early 2010, Congress was shocked—and so was the aerospace industry. On the other hand, supporters of the "New Space" entrepreneurial industry cheered. While the Bush administration had made a few gestures toward a more commercial approach to space, notably with the Commercial Orbital Transportation System (COTS) contracts which allowed private firms to build resupply vehicles for the International Space Station (ISS), by killing the Constellation program the new Obama policy would, in theory, displace the old NASA business model.

Things did not quite work out that way. Congress insisted on keeping the Orion program going and revived the Ares V as the national Space Launch System (SLS), sometimes referred to as the "Senate Launch System."

For the military, the great virtue of Obama's move was to open up the way for a new set of contractors, with lower overhead and, thus, lower costs. The most important of these new firms was Elon Musk's SpaceX, which soon achieved Musk's goal of being able to fly national security payloads on his Falcon 9 and Falcon Heavy launchers.

There was, however, a negative side to Obama's move. It very nearly cut the guts out of the US large solid-rocket industry. One of the main elements of the US status as a superpower is its ability to build and deploy ICBMs and SLBMs. In 2010, these were the Minuteman III, dating from the early 1960s, and the submarine-launched Trident D-5, dating from the early 1980s. Naturally, these had been improved over time and refurbished, but the industrial knowledge needed to keep their rocket engines working was concentrated in a single firm based in Utah, Morton-Thiokol. It may be impossible to accurately calculate how much extra the Defense Department will have to pay for the new Ground Based Strategic Deterrent (GBSD) ICBM, due to this policy, but it certainly is not zero.

During the Obama years, the Air Force suffered from an unfortunate combination of pressures that resulted in many of the most important space programs being neglected. The delays and cost overruns on the F-35 program may have impacted the Navy and the Marines, but the bulk of the extra costs hit the Air Force. At the same time, the supplemental budgets intended to pay for the post-9/11 War on Terror covered only part of the Air Force's costs—in particular, a maintenance backlog built up. Limited budgets also forced the service to cut back on several advanced technology programs, especially in hypersonics.

These problems created a suspicion that the leadership of the USAF was using so-called "budgetary magic" to shift funds from space programs to things like the F-35, which, naturally enough for an "Air" Force, were of greater interest to the people, military and civilian, at the top of the organization. This suspicion was not confined to a few space experts but spread to Congress. Members of the House, who may lack the grand visions of their Senate counterparts (but who sometimes have the time and the motivation to dig into the details of executive branch activities), found that even if they couldn't prove any malfeasance, the USAF was simply not paying as much attention to space as it said it was.

Indeed, since 2001, the USAF and the Defense Department acquired a remarkable skill in explaining why space was important and why the US had to be militarily dominant in the new high ground. Yet, somehow, they never got around to doing much about it. To say that they gave lip service to military space is to minimize their rhetorical accomplishment. The time and effort the department put into avoiding the serious development of US military spacepower, especially weapons that would challenge the sanctuary doctrine, was an amazing feat of bureaucratic legerdemain. Historically comparable, perhaps, to the

British Army's horse cavalry advocates' success in derailing mechanization in the early to mid-1930s.

By the time Trump was elected in 2016, there was, inside Congress, a small, bipartisan band of members who knew that something was wrong and was ready to demand fundamental institutional reform.

Like most post-Sputnik presidents, Trump came into office with few clear ideas about space policy. His lack of interest in space might be best characterized by the fact that, as far as one can tell, he never attacked the International Space Station program as a US giveaway to foreign interests. He was, however, determined to shake things up, to "drain the swamp."

Trump's election may represent one of those "break the mold" moments in American political history, comparable to the elections of Andrew Jackson, Abraham Lincoln, Woodrow Wilson, and Ronald Reagan. It will take decades to know if 2016 was just a blip or was a true turning point. In any case, Trump's slogan, "Make America Great Again," evoked one of the greatest achievements of the nation, the Apollo Moon landings. The new president may not have been too interested in the details, but he was open to doing something big in space.

One promise he did make during the campaign was to reestablish the National Space Council and to put Mike Pence in charge. As a tool of presidential authority, the council has had a mixed record. When it was called the National Aeronautics and Space Council, LBJ used it to push his own enthusiasm for civilian space exploration. Later, vice president Spiro Agnew tried and failed to convince Nixon to propose an ambitious space program that would have included a space station, a base on the Moon, and a trip to Mars. George H. W. Bush later reestablished it as the National Space Council. Trump found it useful as a venue in which to push his various civil space initiatives, but the Space Force idea originated with Congress.

Sort of.

When it was set up, few people on the council—or, for that matter, elsewhere in government—imagined that by the end of Trump's term

America would have a new military service called the Space Force. At best, they seem to have hoped for improvements in procurement that would mean fewer delays and cost overruns on critical space programs. Instead, Trump and Pence put together the most ambitious space program since LBJ. In many ways, it is even more all-encompassing than anything Johnson ever came up with.

In 2017, Congressmen Mike Rogers (R-AL) and Jim Cooper (D-TN) proposed a Space Corps that would be, to the Air Force, something like the Marines were to the Navy. The military establishment was not too thrilled by this idea. The secretary of defense, Jim Mattis, tried to pour cold water on the idea by asking if it would just be another bureaucracy, while former presidential candidate John McCain, chairman of the Senate Armed Services Committee, sabotaged the proposal. According to one story, he agreed to support the Space Corps idea in exchange for the GOP-controlled House supporting his ideas on cyberwarfare. As the story goes, McCain got what he wanted on cyber and then reneged on his promise to support the Space Corps.

Another story tells of a visit by Air Force secretary Heather Wilson and USAF chief of staff Dave Goldfein to Rogers's office that ended in an ugly screaming match. The USAF launched an all-out lobbying effort to kill the Space Corps idea.

By early 2018, the White House and Space Force supporters in Congress had come to a consensus that they would make an all-out effort to include legislation establishing the USSF in the 2019 National Defense Authorization Act.

CHAPTER 3

LEADERSHIP AND PERSONNEL: THE GREATEST CHALLENGE

"Sure, enlisting in the Space Force sounds glamorous, but instead of zapping Martians with your laser blaster you'll probably end up peeling potatoes in the space mess hall."

—David Burge (@iowahawkblog),
March 15, 2018, on Twitter

On March 31, 1942, General Alan Brooke, chief of Britain's Imperial General Staff, wrote in his diary, "It is all desperately depressing. Furthermore it is made worse by the lack of good military commanders. Half our Corps and Divisional Commanders are totally unfit for their appointments, and yet if I were to sack them I could find no better!" This note expresses the unfortunate fact that senior military leadership in war is a job that few men (and women) are able to handle successfully. Even a lifetime of training and experience cannot prepare a leader to pass the supreme test of war. The history books are filled with the stories of men who either failed or succeeded at a horrendous cost.

It has been said that in the Army, officers salute their troops and say, "The enemy is over there; give 'em hell, men." In the Air Force, the enlisted men salute the officers and say, "The enemy is over there; give 'em hell, sir." In the Space Force, we may expect that officers and enlisted troops will salute their computer terminals and say, "The enemy is over there; give 'em hell, robots."

Tobias Naegele, editor in chief of *Air Force Magazine*, in an editorial denouncing the "Space Force zealots," claimed that "the Joint Chiefs will not become wiser with the addition of an eighth four-star general." One could ask if adding the vice chairman of the Joint Chiefs or the commandant of the Marine Corps added any wisdom. Putting a new four-star Space Force commander on the JCS might indeed have a beneficial effect, depending on the man or woman.

The leaders of the Space Force will at some point be tested, either by a crisis or by war or by both at the same time. How they respond and if their response is successful will not only depend on how well they were trained and how good the men and women they command are at their jobs (not to mention the possession of the right technological tools), but also on those intangible elements—charisma, instinct, luck, "presence," humor, and empathy—that make up what is sometimes called "military genius." Finding, cultivating, and tolerating the right people is going to be the hardest job the initial leaders of the Space Force will have.

Howard Bloom, in his book *The Mohammed Code*, has written about something he calls "the founder effect": how the founder of any institution, nation, or religion affects the profound nature of the thing that is founded. Americans are more aware of this than most people due to the nature of the men who used to be called the Founding Fathers. George Washington, in particular, left the nation with a sense that a great leader was one who accepted limits on his power and strived to cultivate virtue in himself and, by example, in those around him.

Dwight Eisenhower created in NATO a multinational alliance whose members accepted US leadership because they knew that the US could be counted on to respect their interests as well as their sense of national pride. The military institution he founded has successfully survived all these decades, in spite of all the efforts to undermine it, thanks to the enduring institutional habits Ike was able to inculcate right from the start.

Selection of the leaders of the US Space Force, both civilian and military, will be critical. The two individuals, the first secretary of space and the first Space Force chief of space operations (CSO) will shape the new organization in ways that will last for decades, if not centuries.

Just as George Washington put his indelible stamp on the US Army, John Paul Jones gave the US Navy its core traditions and Hap Arnold was more responsible than anyone else for the Air Force America has today. The person chosen to be the first military leader of the Space Force will determine not only its future but will also have a major say in the future of the whole national space enterprise, including the civil and commercial aspects of our off-world activities.

It will simply not be enough to find a competent and proven military leader; what is needed is someone with guts, vision, high intelligence, and an open mind. Unfortunately, John "Jay" Raymond, the first CSO, has a reputation for having none of those characteristics. Instead, president Trump and the secretary of defense chose a general well known for "playing well with others." Unless Raymond can surprise his critics or a new generation of space leaders emerges, the Space Force may be doomed to be a minor and not very powerful institution. As of now, the CSO is taking "guidance" from the other services, especially the Air Force, as well as from Congress, non-military bureaucracies, and non-governmental organizations (NGOs) that may not be in favor of America being the world's dominant space power. Raymond may proclaim that his organization will be characterized by "agility, innovation, and boldness," but as of now, we see very little evidence of anything truly bold in Space Force policy, doctrine, or behavior. If Raymond or any other leader were to go boldly anywhere, the powers that be inside the beltway would try and slap him down in less than a nanosecond.

When trying to sum up Prussian general and military theorist Carl von Clausewitz's career, American cultural historian and war researcher Peter Paret wrote, "A passionate soldier comes to regard war as an activity from which assumptions, convention, doctrine must be stripped, so that war can be studied and understood as an expression of social and political life, as a compound of violence, creative genius, and reason." Such leaders are never going to be easy to get along with, but they can be expected to perform under stress. Sadly, few senior US military officers today are truly "passionate soldiers."

As of now it looks like the Space Force is having a hard time finding even a few senior officers that are up to the job. During the Air Force's lobbying campaign against the Space Corps, one retired general is supposed to have said, "You can't put space people in charge of a Space Corps. Have you seen the kind of people we send there?" It looks as if he had a point, though it might be more accurate to have asked, "Do you see the kind of people we promote to flag rank?" Fixing the quality problem at the top of the Space Force is going to have to be a priority for the White House.

Raymond looks like he is having a hard time recruiting a senior staff who're both qualified and, even more important, dedicated to the mission. It was never going to be easy finding these officers. There are few enough of them on active duty, and since the Air Force determined that the military space career field would include officers who have served as missileers operating the nation's ICBMs, the service has made it even more difficult for those who've specialized in military space to the exclusion of all else to get promoted.

Of course, there have been a few missile people who have proved that they are for-real, unadulterated space experts, but these are few and far between. The same can be said about USAF pilots who've forsaken the one true Air Force religion for the near-pariah status of space enthusiast. Getting these people into the USSF senior staff should be a priority, even if it means looking outside of the active-duty officer corps.

Ideally, the original Space Force leadership should include 20 or 30 percent officers from outside the Air Force, not just Army Signal Corps experts who've used military satellites all their careers or Naval officers with long experience using all sorts of space systems, but also Marines and especially Coast Guardsmen who can handle the "Space Guard" aspect of the USSF mission.

In its first few years of existence, the Space Force will have to improvise a whole new style of military leadership. Ideally it will be focused on professionalism, intellectual excellence, and a strong sense of duty and patriotism. The first of these traits, professionalism, is a natural part of the makeup of all twenty-first-century US military leaders. It not only implies that the officers have been properly trained and socialized into their roles, but that they will display the kind of self-control appropriate to someone who carries great responsibilities and may be expected to carry even greater burdens in the future.

The second characteristic is intellectual excellence. This means that the leaders of the Space Force will be, one hopes, relentlessly curious— that they will always be seeking to improve their knowledge of the universe in both its scientific and human aspects. This cannot simply be a matter of making one's way through reading lists provided by the command authorities, even though these are an essential aspect of creating organizational cohesion. Officers must be encouraged to read and study things that may be outside their normal "lanes." The USSF should, as a matter of policy, provide time (not time off) for its personnel to study. This could be occasionally giving individuals a week or so to spend on a study project of their own devising, or it could mean that small groups of Space Force officers go off somewhere for a "reading party" under the loose supervision of a teacher or senior officer.

Most importantly, the leaders of the USSF will all share a strong sense of duty and patriotism. This is critical. All too often over the last decade or so we've seen evidence that the anti-Americanism that has rotted our major universities has seeped into the US military's educational institutions. The infamous West Point "Communist Cadet" is only one example. The reports I've heard from Air Force staff colleges and the personnel decisions made at such venerable schools as the Naval War College indicate that a significant number of officers who have emerged from these places cannot be depended on to support US national interests when they conflict with what are sometimes termed "globalist" desiderata.

There is nothing wrong with exposing military officers to hostile ideas. Unfortunately, what seems to have happened is that anti-American elements have replaced or driven underground pro-US ones and now prevent what used to be considered noncontroversial ideas, such as the concept that, on the whole and in historical context, the US has been a force for good in world affairs.

Instead, we have a generation of young officers who've been brought up reading the communist Howard Zinn's *A People's History of the United States*. Their exposure to alternatives, such as the traditional liberal Samuel Eliot Morison's *The Oxford History of the American People* has been nil. This means that those responsible for training the junior Space Force leaders will have to work hard at "deprogramming" their junior officers before they can be trusted to operate the nation's vital space systems.

One of the great dangers, which we already see happening, is that left-wingers will seize critical positions inside the Space Force's graduate education establishments. Professorships in space strategy are being given to unqualified individuals associated with the Secure World Foundation, which has long been devoted to undermining America's spacepower.

Institutions such as the Air University's School of Advanced Air and Space Studies have been critical to the career paths of ambitious space officers. Giving leftists important positions inside these institutions is allowing hostile powers to determine the makeup of the USSF's future senior leadership.

The Space Force will eventually need its own Space Force Academy, and the choice for the first commandant will be almost as important as the choice for the first USSF chief of staff. Whoever is chosen for this job must make a commitment to stay on for at least nine years (three tours) in order to ensure continuity and so that he or she will truly be able to mold the school into a unique center of excellence. The existing Air Force Academy is adding space courses to its curriculum, which it should have done years ago, but in the long term a separate Space Force Academy is essential and inevitable.

One aspect the Space Force leaders might want to consider is to make the academy only accessible to men and women who have already served two or more years in the Space Force enlisted ranks. This may not ensure that all the future space cadets will be mature and responsible adults, but it will help. The goal is not necessarily to produce junior leaders who will, like Marine Corps leaders, inspire "institutional heroism in the heat of combat," but rather leaders who will consistently perform difficult, technological tasks day in and day out with zero mistakes and will, when needed, show real moral courage.

Of course, physical courage will always be required. Space Force personnel will always have to be ready to fight to defend their terrestrial bases and assets. This means that physical fitness and weapons training will be an essential part of Space Force training. However, leaders must be careful not to allow such combat training to crowd out the main space operations and space dominance missions.

Recruiting and training Space Force enlisted personnel may prove easier than recruiting and training the Space Force officer corps. A young man or woman in high school who wants to join the USSF should be able to look forward to at least four years of excellent, cutting-edge engineering and scientific education—as well as invaluable, hands-on technical experience. (This would be a better education than is offered by most universities, without the worry and burden of dealing with student loans). The USSF will not need more than one or two thousand new recruits a year. It will have few problems finding and signing up some of the cream of the nations' youth.

In early November 2018, an Air Force one-star general announced that the USAF planned to keep control of the Undergraduate Space Training program at Vandenberg Air Force Base in California, at least until the Space Force is well established as a separate service. It will be interesting to see just how much effort and cash the USAF puts into this program knowing that it eventually will have to hand over control to the USSF. If the Air Force does improve the facilities and assigns some of its best people to this training, it will be a sign that the separation is going to be a friendly one.

No matter what, Space Force basic training will have to be rigorous and challenging. It cannot blindly copy the basic training of the other services, but it cannot ignore their experience either. It should naturally include things like close formation marching and community singing that create and reinforce group cohesion and discipline. It should also include lots of software development and things like welding, metalwork, and basic 3D manufacturing. Physical training and weapons training are naturally enough part of any military organization's training, and in this regard the Space Force should not be an exception. A Space Force recruit should emerge from this training ready for specialized training in one of the main branches of the force.

Training senior enlisted personnel is something the US military has, in recent decades, excelled at. Senior USSF sergeants should be eligible to attend a variety of USSF schools, other military schools, and civilian universities. The ideal model for this is the group of senior sergeants, many of whom have PhDs, who manage the GPS control system from a facility at Peterson AFB in Colorado. These men and women have a well-deserved reputation for excellence; if the Space Force can emulate their example in other areas, such as communications satellite control and early warning (SBIRS), it will achieve the kind of institutional excellence that is to be expected of it.

One promising sign is the announcement that enlisted Space Force promotions will be done by selection boards rather than by written tests. This should encourage the NCOs to show original thinking, "problem-solving skills," and qualifications that go beyond what is required of enlisted personnel in the other services.

However, no matter how well the enlisted force performs, it is the Space Force officer corps that will ultimately determine the success or failure of the service. For decades, people inside the defense establishment have been trying and failing to successfully reform the procurement system. In future years, if the USSF is successfully established, administrations and Congresses will judge it by how well it acquires its assets. If it simply continues the existing pattern of delays and cost overruns, then it will find itself condemned and often ignored.

The Air Force failed to convince either the Bush White House, defense secretary Bob Gates, or Congress that it needed more than 187 F-22s. Based on America's need for global air supremacy, it was a bad decision, but the USAF's record of poor program management probably made it inevitable. The Space Force cannot afford such fiascos. The Space Development Agency (SDA), established in March 2019, has a mandate to avoid traditional procurement practices, and should, in theory, help the new service avoid these disasters.

But if the agency is staffed by men and women who have spent their careers working in the heart of the old system, the SDA will be in danger of repeating the kind of problems we've seen with the SBIRS program or with the NRO's 1990s failed attempt at building the Future Imagery Architecture (FIA). Many of the old program managers and specialists are good people who've been frustrated by a system and regulations that consistently cripple the large-scale "programs of record" that the Defense Department develops and builds.

The SDA might want to consider using something akin to the French DGA (Direction générale de l'armement) procurement agency as a model. The DGA is a corps of military specialists who do nothing but help French industry design, develop, and build the weapons and systems needed by the French armed forces and their foreign clients. With their limited budget, and in spite of a level of political interference which is of similar intensity but institutionally different (but at least as bad) as the one afflicting the US, the DGA has managed to avoid many of the delays and cost problems that keep recurring in the American system.

Getting the incentives and accountability balance right for SDA personnel is not going to be easy. The civilian workforce will, without doubt, be protected by civil service regulations. The rewards they receive will have to be both monetarily generous and intellectually and emotionally satisfying. Giving them the chance to develop technologies at the extreme cutting edge (often called the "bleeding edge" since projects of this kind tend to bleed money) will make working for the SDA particularly attractive.

However, since the SDA will often be assigned to build or modify systems whose technology is well understood, there will be limits as to how many experts, engineers, and scientists will be allowed to work on truly exciting projects. Men and women who are assigned to relatively mundane jobs will have to be carefully motivated and closely supervised. We've learned the hard way that in government bureaucracies, or in any large organization, employees and managers tend to get complacent when dealing with systems they think they understand. The 1986 *Challenger* disaster stands out as a prime example of this problem.

Leading a new organization such as the SDA will require an extraordinary individual, one who is technically qualified, morally courageous, and able to inspire deep trust in both subordinates and superiors—especially in dealing with civilian politicians. The first head of the SDA, Derek Tournear, previously was assistant director of research and development for space within the Office of the Under Secretary of Defense for Research & Engineering and previously held leadership roles in industry. He will have to be sensitive to original ideas that may come from outside the Space Force and the traditional aerospace industry, and if he seeks to simply "get along" inside the bureaucracy, then the organization will probably be doomed to mediocrity.

If the SDA is to perform better than the USMC, it will have to learn to cooperate with the so-called "New Space" industry, a nebulous collection of entrepreneurs, financiers, visionaries, and, in the background, a few billionaires such as Elon Musk and Jeff Bezos. Space start-ups seem to have an even higher failure rate than ordinary tech start-ups.

However, the technology they produce is almost always highly relevant to the needs of the Space Force. The SDA will need to keep a close eye on these activities.

The SDA will inevitably exchange people and ideas with other US government space organizations, particularly NASA, NOAA, and the Office of Commercial Space Transportation (generally known as FAA/AST). The agency's leaders will have to ensure that these exchanges are carefully focused in order to achieve specific goals and not simply to allow people to become better-rounded space professionals. That sort of thing is best left to the training and education side of the Space Force.

The Space Force will have to find ways to integrate scientific advice and scientists into almost everything it does. Recently, many in the political world, mostly Republicans, have lost their trust in "science." This has been developing for a long time. During the Cold War, many prominent scientists aligned themselves with the USSR; more recently, the scientific establishment has positioned itself in support of an environmentalist agenda that looks, to some, like a soft totalitarian's wish list. Ensuring that the scientific advice provided to the USSF is reliable and untainted by politics is going to be one of those uncomfortable issues the leadership would rather avoid.

Anyone who wants to study the problem could do worse than to read C. P. Snow's little 1960 book, based on a lecture at Harvard, entitled *Science and Government*. In the first part, he tells the story of the conflict between Henry Tizard, who is entitled to claim paternity of the UK's vital home defense radar system that saved the nation in the 1940 Battle of Britain, and Fredrick Lindemann, Winston Churchill's irascible chief scientific advisor.

The lessons Snow draws from this conflict should be of interest to the Space Force and to its leaders. First of all, the distinction between "open politics" and "closed politics" is a useful one at a time when there is a tendency to demonize all forms of government secrecy. Any serious observer knows that governments of all types tend to over-classify information. On the subject of secrecy, Snow explains: "I have known men, prudent in other respects who become drunk with it.... It takes a very strong head to keep secrets for years and not go slightly mad. It isn't wise to be advised by anyone slightly mad."[26]

Snow also wanted governments to beware of scientists who are euphoric about "gadgets." Perhaps this may have been yet another manifestation of the old British antipathy to new technology. For Americans, however, it is a worthwhile caution. We do tend to become overenthusiastic about new gadgets and tend to misjudge how quickly they can be developed and how much of a real-world impact they will have. This applies to some of our politicians even more than to our scientists. Al Gore, Newt Gingrich, and Hillary Clinton have all been guilty of this at one time or another.

Yet, as the saying goes, "A man's reach should exceed his grasp." The US does have a record of accomplishing things that seem impossible or incredible at first and yet become commonplace and routine in time. Elon Musk's Falcon series of low-cost, reusable launch vehicles is a good example; the microchip industry and stealth technology are others. An allergy to technological euphoria may be just as harmful as an uncritical embrace of anything new and slightly plausible.

One factor in Space Force personnel management that cannot be overstressed is the need for counterintelligence. US military space operations have always been a target of foreign intelligence services. With the creation of the USSF we can expect that these efforts will intensify. Protecting the service from penetration should be a constant concern.

The two most dangerous types of infiltrators are the Aldrich Ames type, who do it for money, and the Ana Montes type, who do it for ideological reasons. It should not be too hard to catch the first type, since keeping an eye on people's spending habits and lifestyle is easier than ever, thanks to the surveillance technology developed by the online advertising industry. The second type is more dangerous and requires a level of political sensitivity that is all too rare inside the US government's civil service.

Ana Montes was not only the chief analyst for the US military's (Defense Intelligence Agency's) Latin American intelligence operations, but she was also well known as an extreme leftist who sympathized with Cuba's communist dictators and detested "American imperialism." Yet somehow no one ever put these two facts together. It's worth asking why.

One can only speculate, but it seems as Montes's superiors regarded her leftist ideology as perfectly normal and could not imagine that she would act as if she believed it by betraying every single secret she could. Could it be that there were people inside the DIA who sort of agreed with her and, while they were not willing to go as far as she did, were not particularly disturbed by her sympathies?

In a democracy, where partisanship is rampant and where almost every issue is passionately argued, the danger of allowing individual political opinions to disqualify individuals for certain sensitive jobs is obvious. On the other hand, common sense would indicate that people whose politics are fundamentally hostile to the interests of the United States are equally a danger to the nation. Finding the right balance is a difficult job, and one that should not simply be left to lawyers and low-level bureaucrats.

Space Force legislation should give the leaders authority to quickly discharge or fire anyone, including Space Force civil servants, who show, by word or deed, that they harbor ill will toward the US and its Constitution. Such authority cannot be unlimited, and it should be expected that it will only be used in extreme circumstances—but the authority should exist.

Since this authority will inevitably be one of the first things that the left-wing legal establishment will try and overturn, it should be very carefully crafted. This cannot be left to ordinary military or government lawyers. The Defense Department should ask some of the nation's best non-left-wing constitutional law professors and scholars to help write this part of the bill. In all likelihood this issue will end up before the Supreme Court.

Counterintelligence and security are just a few of the legal challenges the Space Force will have to deal with. If the leaders are not careful, national and international legal issues will suck up irreplaceable time and effort. Fortunately, there are lots of good civilian space law experts out there and the USSF should not hesitate to hire them.

If it ever becomes a real threat to America's enemies, the Space Force will come under attack from leftist "lawfare" practitioners. The best way to counter them is to let civilians—especially civilians from outside the government—deal with them. Unfortunately, Defense Department (Judge Advocate General, or JAG) and Justice Department lawyers have a less than stellar record of dealing with these attacks. Lawyers from outside the government, whose political motivations are at least as strong as the motivations of the lawfare lawyers, will be better prepared to deal with this than the usual civil servants with law degrees.

It cannot be stressed too much that the success of the Space Force will depend on the quality of the leadership and the followership of the organization. The people of the Space Force will have to be an elite among the elite. They cannot be chosen in the traditional way the military recruits and promotes its people, nor can they be picked according to the current "meritocratic" system in place at America's Ivy League and other top schools. A new set of criteria or a new way of thinking about human resource management must be established.

These criteria will naturally include an emphasis on technical expertise. All Space Force personnel will be well versed in the principles of the scientific method and will be trained to understand the rules of logic. But more importantly, they must be taught *why* doing their jobs is important—and that means being inculcated with intelligent patriotism.

Diversity training, also known as critical race theory training, as it has existed in the US military over the last thirty years, may indeed have been a factor in the slow degradation of the overall quality of American military leaders. Space Force leadership will, one hopes, find a way to convince members of America's wide variety of religions and races, not to mention sexualities, to live and work together without indulging in the dubious joys of identity politics.

However, the Space Force cannot be expected to solve the Defense Department's many sociological problems, let alone those of the nation as a whole. The best we should expect is that it will avoid adding to them or finding new ways to make things worse.

If the Space Force fails and is reabsorbed into the Air Force, there is a strong potential that the USAF space personnel will suffer a collapse of morale that could last a generation. The signal will be sent that Congress and the existing bureaucracy don't care about military space and indeed are hostile to it when it threatens their "iron rice bowls." Few high-quality young officers will choose to pursue military space careers, and those who do will be constantly tempted to fall into an attitude of self-pitying bitterness, anger, and despair.

Even the best Air Force leaders will have a hard time convincing space cadre officers that they are as highly valued as fighter or bomber pilots. The Air Force will be stuck with a faction of men and women who have an institutional reason not to trust their senior leaders. Now that we have a Space Force, the die has been cast; if it fails, it will be impossible to go back to the old structure. If, for whatever reason, the USSF is abolished, sorting through the wreckage will take time and create vulnerabilities that few people in Washington will want to face up to.

One final point on rank: There was considerable discussion regarding new ranks for Space Force personnel, much as there was when the Royal Air Force was established in 1918. Congressman Dan Crenshaw (D-TX), a former Navy Seal, proposed that the Space Force use Navy ranks. Writing in *The Hill*, retired Air Force general and Space Force opponent David Deptula called that proposal "silly" and seemed to imagine that proponents of a naval rank structure had watched too much *Star Trek*. In a rely to Deptula, published on August 21, 2020, Brent Ziarnick, assistant professor of spacepower at USAF Air Command and Staff College, explained that "if [Deptula] were familiar with the Space Force's actual culture, he would know that naval theory is nearly universally accepted as the essential foundation of spacepower theory." Ziarnick contended the issue was deeper than just a matter of rank and involved the emancipation of the USSF from the Air Force, noting that the effort by some inside the military seraglio to undermine the intentions of the commander in chief and to ensure that the Space Force's activities do nothing to harm the institutional interests of the Air Force.

Finally, in December 2020, the Space Force announced its personnel would be referred to as "guardians," and the following month said its officers would keep the same ranks as in the Air Force while several lower to middle enlisted ranks would be changed.

The fight over ranks barely concealed the fight over doctrine, which itself conceals the fight over strategy and what will be the ultimate priorities of the Space Force as a military institution. The biggest problem is, will the force be focused on supporting the warfighter on Earth? Above all, the "poor bloody infantry?" Or will the Space Force focus on establishing dominance in the Earth-Moon system and eventually, decades from now, in the solar system. The USSF will, by the nature of things, have to do both, but in the end the biggest fights will be over priorities—above all budget priorities.

Congress could help by giving the Space Force a better budget than the one currently proposed. A generous "plus-up" for R&D would be useful, as would a slightly higher budget for personnel.

The first doctrinal document put out by the Space Force says that "military space forces are the warfighters who protect, defend, and project spacepower. They provide support, security, stability, and strategic effects by employing spacepower in, from, and to the space domain. This necessitates close collaboration with the US Government, allies, and partners in accordance with domestic and international law." This last sentence is typical of the kind of mentality that could cripple America's military space operations. As a matter of practice, "international law" often depends on who decides exactly what it is and what it means.

PART TWO

FORCE MISSIONS

CHAPTER 4

SPACE COMMUNICATIONS: KEEPING EVERYTHING CONNECTED

The Romans had roads, the Incas had relay runners, Spain had caravels and galleons, the British Empire had steamships and telegraphs, and from the 1960s onward the US has had military satellite communications systems. In each case, the ability to gather information in a relatively secure and timely fashion and to reliably transmit orders has been the foundation of any long-lived regional or global power. When this ability breaks down or is overwhelmed, that power falls apart.

The idea of communications satellites dates back to renowned science-fiction writer and futurist Arthur C. Clarke's October 1945 paper "Can Rocket Stations Give Worldwide Radio Coverage?" In it he described the way broadcasts from a space station could "be provided with receiving and transmitting equipment (the problem of power will be discussed later) and could act as a repeater to relay transmissions between any two points on the hemisphere beneath, using any frequency that will penetrate the ionosphere."[27]

At this time, the Nazi V-2 rockets had demonstrated that space flight was a real possibility. In the Soviet Union, communist leaders were fairly quick to grasp the military potential for long-range rockets. In the US, a few senior officers (such as Hap Arnold) understood this, but postwar budget cuts and the complacency created by America's short-lived nuclear monopoly meant that, aside from allowing the von Braun team to continue dabbling in rocketry, they did practically nothing in this area.

Some interesting progress was made by the US military when in January 1946 the Army Signal Corps bounced a burst of radio energy off the surface of the Moon. The project, named Diana for the Roman goddess of the Moon, "proved that humans could communicate electronically through the ionosphere into outer space."[28] This experiment was an exception; in general, due to the drastic post-1945 budget cuts, the US military stopped most of its research and development work not directly connected to either nuclear weapons or aviation.

The communist leaders in Moscow, however, refused to prioritize postwar rebuilding and devoted their efforts to an all-out drive to catch up with the US in all areas of military power.

After the first Soviet atomic test in 1949, and especially after the Sputnik launch in 1957, all this changed. The US had been challenged to an arms race and, in parallel, to a space race. Fortunately for Washington, US resources were far, far greater than those of its communist foe. By the time the USSR collapsed in 1991, the technological and economic superiority of America and its allies was blindingly evident.

The military side of the space race was dominated by spy satellites, but in these early days the need for secure global communications in order to provide command and control of nuclear forces was evident. At the time, no one understood this better than President Eisenhower. The doctrine of "massive retaliation" depended for its credibility on the ability of the president, (or, as the euphemism put it, the National Command Authority), to give orders to the Air Force's Strategic Air Command and, beginning in 1960, to the Navy's ballistic missile submarines.[29] The communications equipment, largely teletypes, that had been adequate in 1945 was dangerously obsolete.

Beginning in 1958, the military began work on a variety of communications satellites. The SCORE (Signal Communications by Orbiting RElay) program showed that satellite communications was theoretically possible. In 1960, the Air Force launched the Courier 1B into low Earth orbit. Equipped with a transceiver, it proved that a practical space system could indeed be used to transmit voice and text over long distances. By the mid-'60s, both commercial and military satellite communications systems had become a reality.

The Syncom 2 satellite, originally intended for civil communications, was used for military experiments in 1963. A few years after, the Navy began to install satellite terminals on its ships. These were intended for "fleet flagships, carriers, and special command/control ships."[30]

The Navy found that satellite communications changed the way the fleet operated. DSCS-1 (pronounced "discus"), first launched in 1966 and was the cornerstone of America's first comprehensive, dedicated military satellite communications system. At first these satellites seemed extraordinarily valuable; their use was reserved for the most important messages sent by the highest-ranking officials. As the capacity expanded, thanks to the DSCS-2 series (first launched in 1971) and the DSCS-3 series (which began operations in 1982), users from further down the chain of command gained access to the network. The system was seen as "an important part of the comprehensive plan to support globally distributed military users on the ground, at sea, or in the air."[31] By 2001–02, US forces in Afghanistan were using DSCS as well as leased commercial satcom transponders, for "reachback." This concept kept traditional military bureaucratic operations, such as payroll and accounting, logistics bookkeeping, personnel records, and much else to be performed in the US rather than in the theater of operations, saving money and possibly lives. One thing "reachback" proved was that satellite communications was not a command tool reserved for the elite.

In 1965, the Johnson administration decided to try and force America's allies into using the US military's global satellite communications system. This policy was a part of that administration's overall strategy of slowing down US and allied military technology development programs: "The United States *should refrain* from providing assistance to other countries which would significantly promote, stimulate, or encourage proliferation of communications satellite systems."[32] Of course, we made an exception for Great Britain, whose initial Skynet comsats were built and launched with considerable US assistance. This Johnson/McNamara-era policy backfired, especially with de Gaulle's France, whose independent nuclear force was already an object of contention with LBJ. The French government was and still is highly sensitive to the political prestige associated with space technology. The harm done by Johnson's administration to US-French relations in this area, as in so many others, has endured to this day.

In the late 1950s and the early '60s, the question of using communications satellites for nuclear command and control hardly came up. Under the massive retaliation doctrine, the president would give the order to launch an all-out nuclear strike either when he was satisfied that the US and/or its allies were under attack or after an attack had struck home. The communications links were, for the most, part civilian. These were controlled by the then-telephone monopoly AT&T.

This made a certain amount of sense. The landlines already existed and hardly needed much modification to be suitable for the simple task of transmitting the "go" order. At least this was the doctrine. In reality, President Eisenhower, who thoroughly understood the strengths and weaknesses of both the intelligence and early warning systems and those of the nuclear force itself largely embodied in the Strategic Air Command (SAC), planned to manage a nuclear war based on the facts as they might exist at the time.

Ike believed that the massive retaliation doctrine's whole reason for being was to deter the USSR at the lowest cost possible to the US economy. He was also convinced that the Navy's Polaris missile-carrying submarines were the ultimate "ace in the hole." In order for them to be effective, they needed a way to receive orders. Deterrence was one thing; nuclear warfighting was something else, and if a nuclear war actually broke out, Eisenhower may have been ready to throw the operational plans in the trash the same way the US disposed of its war plans after the Japanese attack on Pearl Harbor.

In any case, the question of how to keep control of nuclear forces was—and still is—a difficult and dangerous one. By the mid-1970s, the Air Force hoped that the Milstar series of communications satellites, operating in the extremely high frequency (EHF) band, would provide a solution. At the time, these were probably the most advanced communications satellites ever built. They certainly were the most secure; their cryptographic packages were the best that money and the NSA could provide.

In recent years, the Milstars were replaced by the advanced extremely high frequency (AEHF) series. These were supposed to be replaced or supplemented by the Transformational Satellite Communications System (TSAT), which was cancelled in 2009 due to cost overruns, and also perhaps because the program was too closely identified with Donald Rumsfeld.

Some of America's closest allies, such as Australia, Canada, Holland, and the UK, have access to AEHF services; the primary user is the president of the United States. The system is a huge improvement over the older Milstars. For example, when Bill Clinton wanted to transmit an order for a Tomahawk missile strike to the Navy using the Milstar system, it took about a hundred seconds to send; using the AEHF, President Trump could have sent the same order in 0.03 seconds. The speed advantage may not seem important, but it makes things much harder for anyone trying to intercept or jam the transmission.[33, 34]

Since 2010, America has launched six AEHF satellites, the last one in March 2020 on an Atlas V 551 rocket. In spite of their excellent capabilities, the AEHFs need to be replaced, and the DoD has finally agreed to initiate the new Evolved Strategic Satcom (ESS) program. The design, construction, launch, and operation of this system will be a closely watched challenge for the US Space Force. Though presidents use a variety of military satcom systems, the AEHF is, by far, the most important one. Future presidents will expect ever more secure, reliable, and speedy global communications. They will also expect the system to be survivable. If anything goes wrong with the program, the president (whoever he or she is) will take it personally, and the USSF leadership will not be happy.

For many years, the DoD plodded along with its Defense Satellite Communications System (DSCS) series. It was succeeded by the Wideband Global SATCOM (WGS). Originally called the Wideband Gapfiller SATCOM, this system in its various versions has replaced the DSCS system. The first was launched in 2007; as of now, there are ten WGS satellites in orbit, the most recent one having been launched in 2019. The next one, WGS-11, built by Boeing and which the company claims will be far better than previous models, is planned to become operational sometime in 2023.

In the UHF band the US now has four operational Mobile User Objective System spacecraft (and one spare) in geosynchronous equatorial orbit (GEO). These provide communications services directly to units on the ground such as special forces.

Military demand for satellite communications just keeps growing. In the early years of the War on Terror, the demand came mostly from the unmanned aerial vehicle (UAV) sector. As more platforms become unmanned (such as the Navy's plans for unmanned ships and underwater craft), the need will keep expanding. In the past, some of this demand could be met by bandwidth compression, i.e., squeezing more data into a transmission. We may have reached the limits of this technology, at least for the moment. Quantum computing may provide an answer, but it's impossible to say when real, proven quantum communications systems will become available.

China and other potential adversaries believe that satellite communications systems are an American Achilles's heel. They may be right, but the wide variety of space-based and space-enabled military communications systems make this a hard set of targets to destroy. There is also the problem of US reaction. Even with a direct and well-understood deterrence factor, an attack on these systems might push the

US toward a disproportionate response that might endanger an adversary's vital interests.

The destruction of the AEHF system might be taken personally by a US president who regards it as his or her personal link to the military and to important allies. No US commander in chief, even one who regards him- or herself as a "global citizen," could tolerate such a humiliation.

Over the next decade we are going to see new generations of regular military communications satellite systems—large, complex, state-of-the-art satellites and their associated ground elements—built by the major defense contractors. We are also going to see constellations of small or tiny networked satellites such as the SpaceX Starlink system, which is quickly becoming operational.

America's military communications may, in the end, prove not to be the sitting duck that the Chinese and others imagine them to be.

From a different angle, the unclassified Space Force Capstone doctrine (published in August 2020) describes "space electromagnetic warfare" as "knowledge of spectrum awareness, maneuver within the electromagnetic spectrum and non-kinetic fires within the spectrum to deny adversary use of vital links. Skills to manipulate physical access to communications pathways and awareness of how those pathways contribute to enemy advantage."

Behind this typical example of "Pentagonese" is the not-so-hidden threat that in any conflict, the US Space Force will launch a powerful and determined attack on the enemy's communications networks. It could be that America's long, expensive, and occasionally difficult experience with military space communications gives it an advantage when it comes to understanding how to disrupt or destroy a foe's systems. So the very size and complexity of the US MilSatCom array provides a certain level of deterrence. Certainly any adversary must consider the possibility that a strike on the US system might fail, but that the US reply might be far more devastating than expected.

CHAPTER 5

GPS AND ITS ENEMIES

Since the beginning of human civilization, quality navigation has been a basic instrument of political, economic, and military power. Knowing where you are, where you are going, and how to get to where you are going are fundamental things that we moderns often take for granted. Before the Global Positioning System was deployed, people used things like road maps, distributed for free by oil companies, and road signs. Today, when maps and compasses are built into cars and personal phones, we have even less reason to think about how we are going to go from A to B.

It is difficult, but not impossible, to draw a straight-line connection between the science of navigation and its offshoots (such as cartography and national and imperial power). Medieval European navigators using portolan charts moved around the coasts and the Mediterranean with a fair degree of regularity. The great voyages of discovery, such as those of Vasco da Gama and Christopher Columbus, were less calculated risks than wild and dangerous adventures. It may be politically incorrect to mention it, but they did open the way for five hundred years of Western civilizational supremacy, which is now morphing into a global civilization whether we like it or not.

Over the centuries, sailors, surveyors, and others became skilled in the use of compass, sextant, and other instruments to guide themselves and to make maps. With the invention of radio, a new series of navigational devices was invented and put into use. In the 1930s in both the

US and in Germany, electronic beacons were installed to facilitate civil aviation. Naturally enough, during World War II this technology was put to use by the belligerents, above all to guide bombers.

After 1945, the huge expansion of civil aviation required an equally huge expansion of aids to navigation. Long-range navigation (LORAN), in all its various forms, became the world standard.

For military purposes, and especially in the case of nuclear war, LORAN was inadequate. In the 1950s and '60s, land-based missiles could be fired from sites that had been carefully surveyed beforehand, thus making their accuracy entirely dependent on the quality of their onboard navigation systems. Bombers and ship- and submarine-based missiles needed something different and better. The US Navy's Polaris missile submarines were, as Eisenhower supposedly told JFK just before Inauguration Day in 1961, America's ace in the hole. Yet if these missiles had to depend entirely on the navigation systems available to a submarine which had to spend almost all its time underwater, the accuracy of the missiles would be questionable. What was needed was a way to determine location in a way that could be done quickly, and just by sticking an antenna out of the water for a few seconds.

The answer was a relatively rudimentary satellite navigation system called Transit. The first successful test satellite was launched in April 1960, and the system as a whole became operational in 1964. Twenty-eight of these were launched and operated until they were phased out in 1996. However, the Department of Defense needed a system that could be adapted to the needs of all the services.

GPS traces its origins to the Naval Research Laboratory's Timation experimental satellites launched in 1967 and 1969. These gave the US military a taste of what a truly advanced satellite navigation system could do. The idea soon spread beyond the Navy, and by 1973 the full Navstar program was approved by DoD senior leaders as well as by some civilian bureaucracies. It was easy for the Army and Navy to agree in this case, since the Air Force would be paying for the system and taking all the technical and political risks.

The Global Positioning System may be controlled by the US, but it is most emphatically *global*. Its future role and the role America's allies and friends play in the future of the system will be one of the USSF's major concerns for at least the next few decades.

When the USSR launched Sputnik in October 1957, one effect was to show that by accurately measuring the time it took for the beeping signal to go from the orbiting satellite to receivers on Earth, space-based navigation was possible. By sending up an orbiting transmitter, Soviet Russia achieved a great propaganda victory and, as so often occurs, gave its Cold War foe the reason it needed to begin serious work on technologies that, over the decades, would enhance US global power.

The transmission of radio signals from orbit to Earth had long been discussed (see chapter 4), but Sputnik gave the world a concrete example that could be examined and emulated. The US military was quick to see the advantages for navigation; however, it was not until the late 1980s that the civilian use of space-based navigation signals began to be widely appreciated. At first this came about slowly, but by the end of the 1990s and in the early 2000s, GPS was so thoroughly integrated into the world economy that some leaders, particularly French president Jacques Chirac, would come down with bad cases of "GPS envy."

In any case, the system was soon an indispensable part of the world aviation industry.

The September 1983 shootdown of Korean Air Lines flight 007 by a Soviet fighter led to President Reagan's decision to allow a civil navigation signal to be incorporated into the GPS/Navstar system. Reagan wanted to ensure that the kind of problem that led up to the downing of the Korean 747 was never repeated.

The original GPS/Navstar system was a creature of the Cold War. The spacecraft incorporated a sensor that could detect and locate nuclear explosions. Having a couple of dozen or more of these sensors available would give the US National Command Authority (i.e., the president or whoever would give the orders to conduct nuclear warfare) a way to tell how the actual war was going—that is, if the satellites, the communications links, and the president or his successor survived. The sensors might also be useful as a way to detect clandestine atmospheric nuclear tests, such as the one rumored to have been conducted jointly by Israel and South Africa in 1979 in the South Atlantic.

During the Clinton era, the 1996 TWA jetliner crash off Long Island helped convince the administration to order the military to turn off the Selective Availability (SA) anti-spoofing mode which deliberately degraded the GPS civilian signal. The US government knew that doing this would enable bad actors, such as terrorists, to more effectively use the system for their own purposes—but it figured that the increased level of safety and reliability for civilian users was worth the price. It is known that the terrorists who hijacked Flight 93 on September 11, 2001, used a commercial GPS receiver to back up the systems on the plane they took over.

However, once SA was turned off, a huge industry of new GPS-enabled devices emerged. One technology that has made giant strides is so-called "precision agriculture," sometimes called GPS farming, which has led to a reduction in the use of herbicides, pesticides, and fertilizer in both the developed and, more recently, the developing nations.

In 2004, the George W. Bush administration published a presidential directive that laid out US policy goals for the GPS system:

+ Strengthen and maintain United States national security.
+ Encourage acceptance and integration of GPS into peaceful, civil, commercial, and scientific applications worldwide.
+ Encourage private sector investment in and use of United States GPS technologies and services.
+ Cooperate with other governments and international organizations to ensure an appropriate balance between the requirements of international civil, commercial, and scientific uses and international security interests.
+ Advocate the acceptance of GPS and United States Government augmentations as standard for international use.

+ Purchase to the fullest and feasible extent commercially available GPS products and services that meet Untied States Government requirements. No activities that preclude or deter commercial GPS activities except for national security or public safety reasons will be conducted and a permanent interagency GPS executive board jointly chaired by the Departments of Defense and Transportation will manage the GPS and United States Government augmentations."[35]

This directive, parts of which are still in effect, confirmed that the US would push for GPS to become a global good under US control and, naturally enough, that the US government would try to ensure that US industry would be able to take full advantage of US technical supremacy in this area. The fact that the signal was completely public ensured that US industry would find it difficult to compete. Garmin, a successful US multinational headquartered in Kansas, finds it necessary to manufacture many of its products in China. The inability of the US to fully capture the industrial advantages of its leading position in this field is typical of the limits the global economy puts on those nations that find themselves at the "bleeding edge" of technological progress.

The policy statement also refers to "augmentation." There are two kinds of augmentation: The first is when a signal is sent out from a ground-based transmitter, which has the effect of supporting the signals sent from space. The Wide Area Augmentation System (WAAS) is certainly what the Bush people were thinking of when they wrote this document. The other is when a satellite launched and controlled by an ally or friendly nation sends signals that augment the signals sent by the US satellites; Japan's Quasi-Zenith Satellite System (QZSS) is one example, and India's GAGAN system is another.

It was also at this time that the US government and others began to refer to this activity as PNT (Positioning, Navigation, and Timing). The signals from the GPS atomic clocks became a critical part of our twenty-first-century economy. As one expert put it, "All cell phones,

ATMs, global electronic financial transactions, and the internet run off of the GPS timing function."

The weakness of the GPS signal means that, at least in the civil version, it's pretty easy to jam. One story involved a plumber in New Jersey who did not care for the fact that his employer had installed a GPS tracking device on his truck. He bought a simple, Chinese-made jammer powered by the vehicle's cigarette lighter. Whenever he wanted to stop work early, he would plug in the jammer and head for his boat. Unfortunately, this usually involved driving past Newark International Airport, and every time he drove past the airport, the GPS system used by the airliners and air traffic controllers went haywire. Eventually he was tracked down and his jamming adventures came to an end, but this episode did warn everyone involved of just how vulnerable the system could be.

The weak signal may be the main reason why the DoD is very, very sensitive about anything that might interfere with the GPS signals. The "frequency overlay" issue was the cause of a serious dispute between the US and the EU when the Europeans were in the planning stage of their Galileo satellite navigation project. In theory, this should mean that the military GPS signals are also easy to jam, but as far as we know, enemy jamming aimed at US GPS receivers has been impotent. This may mean that the Defense Department has been pretty good at keeping enemy successes out of the press, or it may indicate that the military has developed ways of securing GPS operations, even in a difficult environment. There is also the possibility that major potential foes such as China and Russia have so far refrained from using their full electronic warfare capability.

In 2003, the story was told that Saddam's military had tried to use a Russian-provided GPS jammer against US forces. My source also said that the jammer was blown up by a GPS-guided bomb, probably a Joint Direct Attack Munition (JDAM).

However, one sign that the US military does not imagine that the military signal is immune is the fact that few new precision guided weapons depend purely on GPS. New types of inertial navigation systems are being developed and installed. These may not be as precise as the GPS-guided systems currently in use, but if the US military comes up against a foe with effective jamming capabilities, we will have alternatives available. The Army, for example, has been moving ahead with its Mounted Assured Position program, which uses a combination of elements to help vehicles defeat GPS jamming and spoofing. Without doubt, the other services have similar programs.

Meanwhile, the ongoing replacement of older satellites by the GPS Block III series had been fairly smooth, at least when it came to the spacecraft. The ground-based command and control and the new devices needed if the military is to take full advantage of the new capabilities (including the M code signal) have been delayed for years, and even the interim products, intended to give the users at least a taste of the improvements, are all behind schedule and also, naturally, over budget.

This is exactly the kind of minor disaster the Space Force is intended to fix; as of now, it appears that the USSF leadership is taking its time addressing the GPS problem. This may just be due to the new organization having to concentrate on getting itself up and running, but Congress is not patient, and if these questions are not resolved soon, USSF leaders will face some uncomfortable questions.

Today, GPS is the global gold standard for PNT services. The USAF had taken steps to keep it that way, but there have been a few stumbles on the way. Competitors such as the European Galileo and Chinese BeiDou systems may be breaking the US monopoly, but as long as the US signal is the most reliable and efficient, America will stay at the top of the satnav food chain. If the Space Force gets it wrong in the future, these alternatives are ready and waiting to grab global PNT leadership from Washington. This would involve a serious reduction in US strategic power.

Yet more and more often, users such as mariners are being warned about interference with the GPS signal. One report says that, "At the beginning of 2020 the Trump administration directed the Department of Homeland Security, the Pentagon and the Commerce Department to minimize use of the GPS system due to the dangers of hacking." It's easy to guess that states hostile to the US have been interfering with the signal in order to reduce trust in America's technology and generally damage the US-dominated world economy.

The dilemma for future GPS programs is how to combine affordability with survivability. For a traditional-style GPS satellite to survive and remain operational in an environment crowded with ASATs, both kinetic and non-kinetic, as well as with lots of fresh and unpredictable orbiting debris, it will have to be armored, equipped with self-defense systems, have a healthy on-board fuel supply for maneuvering, and have a power source that does not include vulnerable solar panels—which probably means a nuclear generator of some sort.

Such a satellite would be expensive to build and to launch. It may be that there are no real alternatives and that the USSF will just have to pay the price for such an "exquisite" spacecraft. One alternative is to build a very large constellation of small (or very small) and very inexpensive satellites. The British government is going to try something like this with its purchase of a 50 percent share in the bankrupt OneWeb commercial satellite communications firm (see chapter 11).

One improvement to the system is embodied in the Air Force Research Laboratory's (AFRL's) experiment satellite, the NTS-3. The most important element researchers want to test is the reprogrammable transmitter, as well as its associated reprogrammable receiver. If this technology works (and especially if this technology works quickly), then we could see a military GPS signal that is far harder to interfere with or jam than is currently the case. It is not impossible that the civilian signals will ultimately benefit from this technology.

Even if it works, the new aspects of the NTS-3 will do nothing to fix the GPS satellites' vulnerability to ASAT attack.

The problem here is that such small and cheap spacecraft could be destroyed by an electromagnetic pulse (EMP) blast or by High-

Powered Microwave (HPM) beams. Protecting them against these sorts of attacks would require shielding, and that would add to the cost and weight of each satellite. One expert claims it is possible to build miniature Faraday cages that would protect these tiny spacecraft from this kind of radiation attack while still allowing them to transmit and receive. If this could be done economically, then swarms of PNT satellites would be viable as the core of a future GPS IV or GPS V system.

Another problem is signal strength; individually, each small satellite would be unable to generate much power. It might be possible to generate a strong signal if hundreds of satellites could be designed to work together. Proving that such a system could be reliable enough for military purposes is going to be tough. Civilian communications constellations, such as Starlink (which is notionally similar to a PNT swarm system), are probably simple to jam and thus would not be much use on a future battlefield.

Finding the right balance and deciding how much risk to accept will not be easy. The ultimate decision will probably fall squarely on a future chief of space operations, and it will not be an easy one. Existing agreements, in theory, limit jamming activities against the military M code and, in theory, also protect two of the civilian signals from interference. In wartime or in a crisis, these agreements would stand little chance of being respected. In a state of what has come to be known as "hybrid war," PNT systems are not immune to interference. Such interference, or spoofing, will probably be deniable and hard to trace.

We can speculate about the way PNT systems were manipulated during the 2020 Nagorno-Karabakh conflict between Armenia and Azerbaijan. It is altogether possible that both sides tried to prevent their enemies from using GPS, GLONASS (Russia's Global Navigation Satellite System) or Galileo to guide their weapons and their UAVs. Some of that interference may have come from Russia or Turkey.

Another future question is how and when to attack non-US PNT systems. In an all-out war, the US will do everything it can to deny the enemy use of all its space systems—including navigation systems. The problem is that in a time of relative "peace," when hostile or semi-hostile states are jamming, hacking, and spoofing America's GPS system, can the Space Force and the rest of the US military strike back with their own cyber weapons? This is a political decision; ideally it would be taken at the highest (i.e., presidential) level.

It's easy to imagine that any plan to launch a cyberattack against a hostile state's satellite navigation system would be questioned and undermined by so-called "deep-state" bureaucrats who might raise the issue of liability for civil damage caused by US countermeasures or even accusations that such cyberattacks would be war crimes. Such opposition would naturally include a campaign of leaks to selected media outlets and congressional staff.

President Trump would almost certainly have chosen to proportionately strike back at any nation attacking GPS. Another president faced with a similar situation might imagine that the political cost was not worth the benefit. In that case, over time the GPS system would cease to be a world standard, and what is now a major national asset would become a burden that would require constant effort to defend it against ever-growing attacks, while other systems would take "market share" from GPS. For the US, this would be an embarrassing and expensive defeat achieved by America's foes at relatively little cost.

For the US government there are two realities it needs to acknowledge. First of all, the GPS system, like all satellite navigation systems, is becoming more vulnerable with every year that passes. Countermeasures are possible but will be expensive. Secondly, the GPS system is essential to the functioning of the world economy and, to some extent, the American military machine and its allies. To say that there are no easy answers to this dilemma is an understatement. The USSF and its leaders are going to have to cope with the need to improve, protect, and maintain the system for the foreseeable future, and this will inevitably absorb money and time.

CHAPTER 6

SPY STUFF

The proper use of intelligence by national political and military leaders is one of those skills that takes years to learn, and even then can confuse and mislead. Very few twentieth-century leaders were able to correctly and effectively use even the best information. For example, thanks to a combination of tradition and communist ideology, Stalin generally had access to excellent intelligence, but as far as we know, in only two cases (Moscow in November and December of 1941, and during the comprehensive effort to steal American nuclear secrets) was he able to transform the flow of secret information into effective strategic action. During its existence, the Soviet Union may have had more than a few intelligence-driven tactical successes, and its counterintelligence efforts were outstanding, but Stalin himself proved easily misled by his suspicious nature and by his amazing trust of Adolf Hitler.

In the West, only Churchill and Eisenhower managed to fully grasp the strengths and limits of their respective intelligence services before they gained national power. Most leaders learn from experience, sometimes at considerable cost, that their intelligence information is often incomplete, flawed, and influenced by the habits, prejudices, and institutional imperatives of the agencies providing the information.

With those facts in mind, any halfway competent political or military leader quickly learns that intelligence in the modern age is the rock on which his or her statesmanship must rest. There is an old saying that "plans are useless; planning is essential," but it could equally be said that "secrets are useless; secrecy is essential."

In the twenty-first century, space-derived intelligence is no longer a luxury available only to a few experts working for the superpowers. Every nation with any sort of pretension to a role in global or regional power politics has built, or is building, spy satellites of one sort or another. This is the reality that the USSF must learn to live with and to deal with, even if it means destroying hostile, threatening spacecraft in orbit.

The Space Force, like the Air Force, will naturally have a role in supporting the launching and operation of America's intelligence-gathering spacecraft. How much of a role is unknown and, unless some major leaks take place, will remain unknown. We know that the military has a major ongoing role in deciding what and when the NRO targets with its intelligence-gathering systems. One thing is almost certain: As the technology for smaller and cheaper intelligence-gathering satellites improves, the demand for tactical, space-derived intelligence will expand. It will be the Space Force, rather than the NRO, which will take the lead in this area.

The NRO will stay focused on providing information for the president and for top government officials. The USSF will concentrate on supporting Army and Marine Corps battalion and company commanders and their equivalents in the Air Force and Navy. The exact division of responsibilities will need to be worked out in detail. Sadly, neither the NRO nor the USSF will find it easy to allow the other side to "win" this bureaucratic battle. But one can always hope for a pleasant surprise.

However, intelligence derived from satellites is only as good as the value added by interpreters and analysts. The skills the US intelligence community has perfected over the decades are, according to reports, awesome. Though there have been more than a few intelligence failures in this area—the Iraqi weapons of mass destruction (WMD) disaster, for example—the NRO and the other agencies have done a remarkable job.

The failure in Iraq is of interest because it shows the limitations of a type of intelligence known as MASINT, which stands for measurement and signature intelligence. This "art" is aimed at figuring out what is happening inside buildings, caves, tunnels, and other places which are normally concealed from spy satellites. It relies on extremely sensitive hyperspectral sensors which operate in hundreds or possibly thousands of separate bands on the electromagnetic spectrum. The data from these sensors is run through giant libraries of signature data to produce results.

It supposedly works like this: The sensors gather data on the smoke emerging from a suspected missile factory in, say, North Korea. The spectral signature of the smoke is then compared with smoke from known missile factories, and the analysts seek to match the signatures to determine if the factory is building missiles and, if so, whether those missiles are being fueled by solid or liquid propellent.

In 2000, NASA launched the EO-1 (Earth Observing-1) satellite, one of the first civilian spacecraft to carry a hyperspectral sensor. As a tool for environmental science, it was extremely useful; it also opened a window on the little-known world of MASINT.

By 2002–03 I assumed that the Bush administration's claims about Iraqi WMD were backed up by MASINT evidence. Obviously, I was wrong; the nerve gas stockpiles did not exist, though Saddam's regime was obviously ready to get back into the chemical warfare business the minute it thought it could get away with it. Still, one can only speculate why the absence of MASINT evidence did not raise doubts inside the White House and the DoD. Alternatively, it is possible that the MASINT experts claimed to have detected Iraqi WMD where none existed.

If, in fact, there was a determination backed up by MASINT that erroneously showed the Iraqis did have nerve gas stockpiles, this news would have almost certainly been leaked. A mistake like that cannot be covered up indefinitely. It might be an understandable mistake to get a reading that showed a nerve gas storage site when in fact it was a warehouse full of pesticides, but if such a blunder occurred, then the public has a right to know. It would seem that after all this time, if the US MASINT analysts made this particular mistake, someone would have made this fact public.

Rumsfeld must have been aware of MASINT technology and of its capabilities and limitations. Other leaders such as Dick Cheney, George Tennant, and Colin Powell must also have been aware that their assessment of Iraq's WMD status lacked this element of confirmation. Did they persist in their belief that there were, in fact, poison gas stockpiles because other evidence was even more persuasive? The reports about the missing stocks of precursor chemicals seemed plausible, even if the reports from defectors such as "Curveball" were not (and apparently could not be checked since German intelligence would not allow the CIA to interview him).

The whole fiasco proved that the intelligence community is as vulnerable to groupthink and peer pressure as any university faculty. Indeed, like many US government institutions, its leaders are far too impressed with academic credentials. Even the best intelligence-gathering technology is useless unless it is analyzed and interpreted by men and women with open minds and moral courage, attributes which bureaucracies by their nature try to stamp out.

All nations—as well as terrorist groups and guerrilla movements—do their best to hide their activities from enemy (and most of the time from friendly) spies. Space-based imagery intelligence systems are well known and well understood. Even a gang of Afghan tribesmen can easily find out when the major US KH (Keyhole) satellites operated by the NRO will be passing overhead and can take cover. More sophisticated states and organizations will set up decoys and employ other deception methods. This cat-and-mouse game goes on all day, every day in spite of peace agreements, cease fires, or the presence or absence of so-called "confidence-building measures." This is one of the reasons why today's leaders cannot have the same level of trust in the products of space-derived overhead imagery that they could back in the 1970s or '80s.

Does this mean that America's huge investment in highly capable imagery satellites is useless? Not really. First of all, every bit of energy that foes and potential foes spend on hiding or trying to hide from our spy satellites is a bit of energy they cannot use against us. There is also the fact that with more and more commercial and allied observation satellites in orbit, the "bad guys" are finding it almost impossible to determine when the skies overhead are clear of sensors. The big unknown is how skillful the image interpreters at the NRO really are, and how good the CIA and military analysts who turn the product the NRO gives them into usable intelligence are.

One sign of the direction in which things are moving is the interest the Space Development Agency is showing in the large-scale purchase of commercial remote-sensing data. Before now, the National Geospatial-Intelligence Agency (NGA) had signed deals with various information suppliers and these proved useful, but the limits of government contracting regulations ensured that the full potential of this new industry was never properly exploited by the US military. Perhaps the SDA, with its more permissive contracting structure, will succeed where the NGA stumbled.

There have been a lot of mergers and acquisitions in the commercial remote-sensing industry. The acquisition of US satellite imagery pioneer DigitalGlobe by the space-focused Maxar conglomerate looks like an example of corporate empire building. One relatively new firm, Planet Labs, looks healthy enough to fend off any corporate raiders, at least for the moment. One of its satellites provided proof that at least two Turkish F-16s were based in Azerbaijan and may have taken part in the 2020 Nagorno-Karabakh war.

As an aside, it's unfortunate that the news media has not made more use of satellite imagery. There seems to be a reluctance throughout the industry to use pictures from space on a regular basis. This may be because the news executives don't understand or trust this news source, or it may be that, like many left-of-center folks, they dislike anything that smacks of military technology. Or it may be that they lack the budgets to pay for the images themselves and, more importantly, the specialized talent needed to vet and interpret pictures taken from orbit.

America's highly capable radar imaging satellites, commonly called the Lacrosse series, can sense what is happening on the Earth's surface in all weather conditions, and at night are amazing "exquisite" pieces of technology. They are also amazingly expensive and amazingly vulnerable. In the early moments of any major war involving the US, they would be destroyed. In normal circumstances, they give our leaders and our analysts data they cannot get anywhere else and that no other nation can match.

Yet they suffer from the same limitations that our other imagery satellites have. They cannot see underground. One of the great future challenges for the NRO and the Space Force is to find ways to detect what is going on underground. It is becoming easier and easier to build structures underground. Elon Musk's Boring Company is on its way to revolutionizing this field, but even before Musk got involved, better machines have been making tunneling much cheaper.

There is a story out of the Middle East that the Israel Defense Forces (IDF) and the Israeli defense industry were frustrated by their inability to detect the attack tunnels that Hamas was digging from the Gaza Strip into southern Israel. Their attempts to use ground-penetrating radar had failed, and nothing seemed to be working until some bright spirit said something to the effect that "the oil industry does a great job detecting anomalies deep underground. Why not look at their technology?" After overcoming some bureaucratic resistance, the IDF tried using methods perfected by the oil exploration industry and soon had located all (or most) of the Hamas tunnels.

The oil industry, of course, has access to the Earth's surface from which to take their seismic readings. To do so from space would take a conceptual and technical breakthrough. However, using lasers to measure minute movements of the Earth may allow analysts to figure out what is happening underground.

The most widely used form of intelligence gathered from space are the various products of America's eavesdropping satellites. These satellites, with code names like Orion and Magnum, usually sit in GEO, where they deploy huge antennas to intercept even the faintest transmissions from Earth and send them to America's intelligence agencies, mostly the NSA, for interpretation and analysis. Twenty years ago, or even earlier, it was evident that the volume of data from these sources was far greater than could be processed by simple humans. The CIA and others developed or procured software tools that could, in theory, mine the data for useful intelligence. While some facts about these systems have emerged, thanks to the Edward Snowden leaks and other revelations, lots of "stuff" continues to remain secret.

One can guess that more and more information processing is taking place onboard the satellites. This prevents unneeded or unwanted data from being sent to Earth-based analysts. At first glance this is not a bad idea; reception antennas are a finite asset, and without some sort of filtration their capacity could easily be overwhelmed. On the other hand, the software that determines what is and is not worthy of transmission could be flawed and thus prevent significant indicators from being examined. The only real answer would be to constantly reexamine the performance of the onboard software and update and modify it on a regular basis. Such activities are not at all cheap or easy.

Yet no matter how good the data and the processing software are, there is still the question of how good the intelligence is that gets into the hands of the nation's decision makers, especially the president. The breakdown of trust between Trump and the intelligence community looked like an unprecedented disaster. Yet conflicts between Republican presidents and the CIA go back to the Nixon administration; certainly Reagan had no reason to think the agency was unequivocally on his side. This problem is not going to go away.

While most of the intelligence community spies on earthly activities, the Space Force will be spying on off-planet activities. This will go well beyond the usual Space Situational Awareness (SSA) operations. Typically, SSA is done by telescopes which track activities in orbit and compare what is seen with a huge catalogue of space objects that previously have been tracked. If something new is detected, analysts can examine it more closely and decide what exactly they are looking at. The US, like other nations, also uses large radar arrays for the same purpose. This ongoing effort has to be supplemented by a new class of inspector satellites which maneuver close to a space object of interest and examine it, visually and electronically. Russia has been using such satellites, and there is no reason to think that the US has not developed similar technology.

From the point of view of whatever nation's spacecraft are being inspected, it may look as if they are about to be attacked. There is little if any difference between the behavior of an inspector satellite and a co-orbital ASAT. This has given the arms control advocates an opening to try, once again, to push for a space code of conduct, hoping to force nations, especially the US, to adhere to an agreement to keep their spacecraft a certain distance from other nations' spacecraft. Since these inspector operations are mostly kept secret, the encounter(s) of France's satellites with Russian inspectors is an exception. It is impossible to know just what has been going on, but based on past experience, if the US were to agree to such a code, Washington would find itself limited in ways that China or Russia would not be.

If US military SSA is not hobbled by international agreements, the Space Force will soon have to start thinking about how it plans to monitor space activities in the Earth-Moon system and beyond—in deep space. As humanity expands into the solar system, military activities will inevitably expand with it. Some of those operations will be out in the open, such as those of the Space Force's own Space Guard. We should expect that other nations will emulate the Space Guard just as they are now slowly beginning to emulate the Space Force. But the most dangerous military space activities will be clandestine. Detecting these will be one of the toughest and possibly most controversial jobs for Space Force intelligence.

Trying to determine if a Chinese probe visiting the asteroid belt is on an exploration mission or is a camouflaged weapon intended to threaten or destroy a US asteroid-mining operation is going to be difficult. Analysts will argue over each detail, and it may be impossible to send a definitive answer to the White House or to Congress. This is one example of the dilemmas the Space Force leaders will find themselves facing and, thanks to the flaws we now see in the advanced military space education establishment, it is likely they will not be trained to deal with them.

CHAPTER 7

MISSILE DEFENSE

Alongside its three principal space support missions—ISR (Intelligence, Surveillance, and Reconnaissance), communications, and PNT—the USSF will find itself under constant pressure to provide America's various missile defense systems with early warning and target tracking data in real time. The Defense Support Program (DSP) and Space-Based Infrared System (SBIRS) satellites and the new Overhead Persistent Infrared (OPIR) have given and will give the missile defense systems the early warning they need. What is needed and does not yet exist is a constellation of satellites to track ballistic, cruise, and hypersonic missiles; their warheads; and their protective decoys in flight. This technology is being developed for the OPIR program, but it will not be operational before 2024. A few experimental satellites might give the Space Force some rudimentary capability before then, but it's not something anyone can count on.

The two Space Tracking and Surveillance System (STSS) satellites launched in 2009 (and first tested in 2010) gave us some idea of how missiles in LEO and in the upper atmosphere could be detected and tracked. Sadly, the Obama administration chose not to follow up this program with a set of fully operational models. Ten years later we are scrambling to put together a program that should be well into its design lifespan. The two original satellites still in orbit are, as far as we know, providing useful tracking data—when they are in the right place at the right time to track things—like North Korean, Iranian, Russian, or Chinese launches. In the years since they were first tested, it is possible that they have had their software upgraded and are now capable of seeing things (such as short-range rockets) that they could not detect ten years ago. Still, there are just two of them and thus are limited.

The US military has made some progress integrating sensors and communications systems with artificial intelligence technology to produce the Advanced Battle Management System (ABMS), which was tested, with only a few glitches, in late August 2020. The test included using a shell from an M109 155 mm howitzer to take down a drone pretending to be a cruise missile. If it works as promised, the ABMS could represent a major step forward in US missile defense, but the DoD's unfortunate record of delay and cost overruns in the software development area must be considered.

The Trump administration failed to make any major changes in US missile defense policy when it presented its long-expected Missile Defense Review (MDR) in 2019. The document presented some useful and carefully edited information about the various threats the US and its allies face from China, North Korea, Iran, and Russia; it also laid out America's cooperative projects with our friends and allies. However, aside from two promises to improve and enhance exiting weapons, it largely maintained the policies the DoD has had in place since Clinton's second term. The department once again successfully resisted pressure from Republicans in Congress to revive the Brilliant Pebbles space-based interceptor program or to press ahead with a serious strengthening of the existing GMD (Ground-based Midcourse Defense) homeland defense system.

The MDR says that "space-basing may increase the overall likelihood of successfully intercepting offensive missiles, reduce the number of US defensive interceptors required to do so, and potentially destroy offensive missiles over the attacker's territory rather than the targeted state. DoD will undertake a new and near-term examination of the conceits and technology for space-based defenses to assess the technological and operational potential of space-basing in the evolving security environment."[36]

As of now, we have not seen any public sign of this assessment, which suggests that the idea has been quietly buried or sent to bureaucratic limbo. The resistance to effective national missile defense within official Washington is still strong.

As of now we have forty-four ground-based interceptors (GBIs)—forty in Alaska and four at Vandenberg just north of Santa Barbara on the California coast. These were originally conceived as part of the Clinton-era national missile defense plan which was taken up and prioritized by the George W. Bush administration. The younger Bush had promised to build a missile defense system, but he had not specified what kind or how big a system he wanted. Surprisingly, neither his vice president, former defense secretary Dick Cheney, nor his secretary of defense, Don Rumsfeld, intervened. They allowed the Clinton plan to proceed almost unmodified.

There were many complaints that the system was rushed and not sufficiently tested before it was declared operational. As valid as the complaints might have been, Bush obviously believed that something was better than nothing, and it was not until 2018 that a full-scale—and very expensive—successful test of the whole system took place.

In any case, missile defense advocates have been pushing to expand the system to include at least twenty more interceptors in Alaska and at least twenty more based on the East Coast at Fort Drum in upstate New York. However, the program was torn apart when former NASA administrator Mike Griffin, head of the Pentagon's R&D activities, ordered the cancellation of the Redesigned Kill Vehicle (RKV) program and slammed the brakes on the build-up of additional interceptors. Griffin may not be the nicest guy in the world, but it must be remembered he is an outstanding, if conservative, engineer. He may have had good reasons to throw a monkey wrench into the program. Sadly, it is at least possible that his action came about because of pressure from those parts of the bureaucracy which are still hostile to national missile defense as an idea and are committed to the MAD doctrine.

The Trump MDR made two relatively significant moves in the direction of providing the American people with a more robust defense than that provided by the forty-four GBIs. He ordered the Missile Defense Agency (MDA) to begin work on transforming the Navy's SM-3 interceptor—in particular the SM-3 IIA version—and the Army's Terminal High Altitude Area Defense (THAAD) system into weapons that could destroy incoming ICBM warheads that were not hit by the GBIs. Missile defense advocates have long pushed for this move. The various models of the SM-3, including the one used for the 2008 satellite shootdown, have proven fairly reliable in tests and equip dozens of US cruisers and destroyers as well as a few Japanese ships and has been ordered by South Korea. When the Obama administration killed the George W. Bush plan to station descaled GBIs in Poland, it substituted SM-3s as part of the Aegis Ashore system, which is now operational in Romania and perhaps by 2022 in Poland as well.

The Japanese plan to install two Aegis Ashore systems to help defend their home islands against North Korean missiles was cancelled due to local opposition and to the amazing combination of ignorance and fear displayed by Japan's military bureaucracy when faced with unexpected obstacles. Japan's government is still struggling to find an effective and affordable way to defend the home islands against missile attack.

The US has not yet decided whether to install Aegis Ashore on Guam or to upgrade the existing installation in Hawaii to make it a full Aegis Ashore system. Even without the capability to intercept ICBMs, this move would considerably enhance America's position in the Pacific and make any attempt by China or anyone else to challenge America's dominance of the central Pacific extremely hazardous.

Testing the SM-3 against an ICBM-style target took place in late 2020. SM-3s could simply be installed on Navy ships and deployed off the US coasts. The Navy would resist this since it already suffers from a shortage of escorts, and the Burke-class destroyers are needed wherever the Navy sends its carrier or amphibious forces. Building a dozen or so Aegis Ashore installations in the continental US as well as in Alaska would be expensive, but such a system combined with an eventual build-up of the GMD system would go a long way toward making the results of an ICBM attack on the US homeland unpredictable. This would help protect the homeland without having to constantly threaten global nuclear annihilation.

The Lockheed Martin THAAD has been under development since the mid-1990s. Its progress was initially slow and painful, with lots of test failures, but eventually the MDA was able to build a reliable system, and the US Army has deployed nine THAAD batteries. Nations such as the United Arab Emirates and Saudi Arabia, who need to defend themselves against the threat of Iran's long-range weapons, have ordered it. The deployment of a single THAAD battery to South Korea caused a major political uproar when China launched a propaganda offensive against the move. It failed, but not before causing a lot of stress to the US-South Korea alliance.

THAAD's capabilities are impressive, so why is the Pentagon unwilling to buy more and to invest in a "deep magazine" program for the existing batteries? One way of getting around this problem is to integrate the THAAD radars and control systems with the Patriot missile. A recent test indicated that the Army could deploy combined THAAD/Patriot batteries that would be more difficult to overwhelm than either system if deployed alone.

The THAAD, like the SM-3, has always had the potential to intercept incoming ICBM warheads in their terminal phase. Lockheed Martin has repeatedly tried to convince the US government to fund such a development program. Neither the Bush nor the Obama administrations were willing to take this step. In the case of Bush, it was probably for budgetary reasons, while for Obama, homeland missile defense was not much of a priority.

Tactical and theater (or operational) missile defense weapons and their associated sensors and control systems have long been easier to sell to DoD policy makers and establishment politicians than "Star Wars" homeland defense. First of all, its technologically easier to shoot down a short-range ballistic missile and/or its warhead than to intercept a weapon with intercontinental range. The shorter the range, the slower the weapon. It also is politically attractive to have weapons that can defend allies. Finally, and perhaps most importantly, tactical and theater defenses do nothing to challenge the MAD doctrine.

The Patriot system is a good example of a weapon that has been in development since the late 1960s (when it was called the SAM-D). It was originally supposed to have some capability against tactical ballistic missiles, but at some point this was excised from the program's requirements. The requirement was restored in the 1980s. The story, as reported, is that senator and future vice president Dan Quayle pressured the Defense Department to restore the system's ability to shoot down short- to medium-range missiles.

During the 1991 Gulf War, the PAC-2 missiles that were available actually did hit many of the Iraqi-modified Scuds that were aimed at Saudi Arabia and the Gulf countries. Their performance, like that of any weapons system when faced with real combat, was imperfect. The Patriot's role in defending Israel was even more "imperfect." What is historically important was not the system's imperfections, but the fact that it worked at all. The 1991 war marked the first time that any ballistic missile of any kind had been shot down in actual combat.

Since then, the system has been constantly upgraded and improved. In particular, the introduction of the PAC-3 hit-to-kill missile gives the system a more reliable way to destroy tactical ballistic missiles. The improved versions of the PAC-2 have shown they have capabilities against cruise missiles as well as other targets. The Israelis used a PAC-2 to shoot down an SA-5 (S-200) missile that had crossed into their airspace.

The international success of the Patriot means that the US has a large base of users whose experiences can be drawn on to decide what kind of improvements need to be made to the system's hardware, its software, and to the way the crews are trained. The downside is that having lots and lots of non-US personnel with access to the system means that the secrets involved are far less secure than they would be if the Patriot were only used by US forces.

Beyond the aforementioned interceptor missile systems lies the realm of directed energy. Back in the 1980s, the opponents of Reagan's Strategic Defense Initiative often tried to confuse the issue by claiming that SDI would consist of lasers and other exotic weapons that seemed right out of science fiction. This ignored the way the program was designed. It was a large-scale set of research programs that included both near-term developments—such as the ones that produced today's GBIs—and longer-term projects that included lasers and other direct-ed-energy programs.

It's been said that "lasers are the weapons of the future and always will be." Lasers may have numerous military uses, such as range-finding, laser radar, and, more recently, dazzling the guidance sensors on heat-seeking missiles, but they have never proved themselves as instruments of destruction. This is changing.

In August 2020, the Israelis released images of a Hamas attack-balloon array being destroyed by a locally built laser weapon. Popping a few balloons may not seem significant, but it was the first time defensive destructive lasers had been used in real-life military operations. The event was a brightly lit signpost along the path to fielding direct-ed-energy weapons.

For decades, some experts have believed that various kinds of battlefield lasers were on the verge of being produced. In the US, billions have been poured into research and development, most notably by the Missile Defense Agency and its predecessor organizations. Israel, which is under constant attack by a variety of projectiles of wildly varying sophistication, has the skills and the motivation to move these weapons out of the laboratory and into combat.

The weapon the Israelis used, Light Blade, was not very powerful, but it did the job. Unfortunately for the inhabitants of the Negev region, there is only one of these systems available at the moment, though it is likely that others are being built and deployed as quickly as possible.

Another Israeli system is supposed to be ready for testing sometime this year. If it lives up to expectations, it will give the Jewish state a new defensive capability against the threat of large salvos of rockets and missiles that Hezbollah has been preparing to launch ever since the end of the Second Lebanon War in 2006.

Meanwhile, the United States has been fitting defensive lasers to some of its ships; these have been tested, and they seem to work against full-size drones. The Navy, however, has been moving slowly, and it will be years before it has the laser weapons it needs to defeat incoming missiles. There is now a clear roadmap to putting these systems on ships.

The US Air Force has been experimenting with lasers that can defend its airfields against small drones. They know just how much effort Russia has had to put into defending its bases in Syria against weaponized commercial quadcopter drones. The Americans are hoping not to be caught unprepared.

Yet there is one area where a balloon-popping laser would certainly be extremely useful—and that is in space-based missile defense.

When it reaches orbit, an ICBM sheds its booster and deploys its nuclear warhead or warheads and also a number of inflatable decoys. As part of the Brilliant Pebbles project, it was proposed to equip the small spacecraft with lasers that would be aimed at both warheads and decoys. If an object wiggled when struck by a laser, then the Brilliant Pebble would recognize it as a decoy and leave it alone; if the object did not wiggle, it obviously must be a warhead and would be targeted for destruction. The concept was logical, but since the whole program was canceled in the early months of the Clinton administration, it was never developed or tested.

Now, with the deployment and use of balloon-popping lasers, the idea of using lasers to detect and destroy decoys in space should be back on the table. At the moment, detecting and tracking decoys is one of the most difficult jobs the GMD national missile defense system faces. A test in 2018 showed that it could be done, but the effort needed to shine a radar on the missile at just the right time in order to see the decoys inflate, and then to track them while they fly in formation with the warhead(s), stretches the system's capabilities to the extreme limit.

Having a constellation of laser satellites in orbit that could "pop" the decoys would help make the mid-course interception problem much easier. Sadly, not much progress has been made in turning GMD into a comprehensive set of missile defense weapons. The effort to build defenses against cruise missiles had, until recently, almost stopped entirely (though there is a single battery near Washington that, under the right circumstances, might defeat a limited cruise missile attack on the capital). The Defense Department's program to build a system to counter hypersonic missiles is now in diapers and will probably remain so for years to come.

The US and Israel are certainly not the only nations developing lasers. In September 2020, Trump's defense secretary Mark Esper pointed out that China and Russia have placed weapons on satellites. Both nations have long been working on laser weapons; obviously, the US believes they have made significant progress.

One concept the Defense Department is thinking about is building an array of sensor satellites that could track hypersonic missiles and provide valid targeting data to interceptors. So far, the DoD has not come top with a plan to build interceptors, but proponents seem to think that if they get the sensors and the software right, they can then build actual weapons that might be able to kill enemy missiles traveling at Mach 5 or more. This seems a tad optimistic. It's probably due more to budgetary pressure than to the belief that a "sensor first" approach is the best way to proceed.

One sign of how seriously the Pentagon takes the hypersonic threat is the quick decision in October 2020 to award contracts to both SpaceX and L3Harris to build prototype satellites to test OPIR technology. Neither of these firms has a reputation for building this type of satellite. SpaceX has never before built a spacecraft for the Defense Department, so the SDA is taking a chance. But, since the larger, traditional contractors would almost certainly take their own sweet time to build such systems, it's a good bet that SpaceX can do the job quicker—and perhaps better—than the big boys.

This project is just part of the effort to replace the SBIRS satellites which are, one must emphasize, performing credibly well, but at a very high cost with new, less "exquisite" systems. These new systems will, if all goes as hoped, combine the early warning and intelligence (battlespace characterization) functions of SBIRS with the tracking and decoy detection function currently carried out by Earth-based radars. The success of such a program will ultimately be measured by a cost-effectiveness metric that probably does not yet exist.

It may be politically impractical, but if the DoD were to build a constellation of small satellites—some with sensors, some with decoy-busting lasers, and some interceptors—all built to a similar design and sharing many of the same basic parts, we could have a robust, affordable, boost-phase missile defense system. Being able to launch dozens of these small satellites at the same time would not only save money, it would complicate any foes' targeting problem.

CHAPTER 8

THE SPACE GUARD: COASTIES IN THE SOLAR SYSTEM

The US Coast Guard came into existence in 1915 with the merger of the old Revenue Cutter Service, which can be traced back to Alexander Hamilton's efforts to collect taxes on the high seas, and the Life-Saving Service. The Coast Guard is a quasi-military service whose principal job is to enforce US laws in US waters. It is well known for its traditional rescue and anti-smuggling missions, but it also enforces maritime safety codes and leads the fight against oceanic pollution. Its wartime missions in support of the Navy are equally famous. Of all the armed services, the Coasties may have the most positive public image—that is, in those rare moments when the public thinks about them at all.

Twenty years ago, it was commonplace for people in Washington who were skeptical of the Space Force concept to say something like, "We'll only need a Space Force when we have settlements and industrial assets in space that need protecting." No one really knows if they were just being cynical or if they were serious. In any case, we are fairly quickly moving toward the point when there will be US settlements on the Moon, and firms are working on ways to mine both the Moon and asteroids. It's only logical to assume that some entrepreneurs will look for ways to process these materials into usable products in space before sending them down to Earth. This is sometimes called the "Downhill Economy."

One of the biggest obstacles to the creation of this new economic domain is the way the 1967 Outer Space Treaty has been interpreted. One US legal scholar wrote that the "appropriation provision of the treaty is arguably unclear and undefined and therefore unworkable. Critics argue that the provision is a result of the socialist ideals that were prevalent at the time but it is outdated and at odds with today's prevailing free-market economy." This "socialist" interpretation has made it difficult for start-ups and even for established enterprises to raise capital or, what is perhaps more important, to buy insurance for space mining, manufacturing, and colonization projects.

Trump's executive order on space commercialization and property rights and the deal he proposed to other spacefaring nations, the Artemis Accords, open the way for private enterprise to begin to operate throughout the solar system. The accords were signed in October 2020 by the US and seven other nations—Australia, Canada, Luxembourg, Japan, Italy, the UK, and the UAE. Other nations are expected to eventually sign on. Once in place, however, some sort of enforcement mechanism will be needed to make sure that all the players stick to the agreed rules. The US mechanism should, and probably will be, the "Space Guard" element of the USSF.[37]

The relevant part of the Artemis Accords reads: "The parties affirm that the extraction and utilization of space resources does not constitute national appropriation under Article II of the Outer Space Treaty. The parties further agree that domestic enforcement of contracts and other legal instruments relating to space resources is not prohibited by the Outer Space Treaty." If and when the Space Guard is created, it will be the obvious agency to handle "domestic enforcement." Just as one of the Coast Guard's roles is to serve as "sea cops," the Space Guard will be the nation's "space cops."

America already has one small but important element of a Space Guard in the Federal Aviation Administration's Office of Commercial Space Transportation, known as FAA/AST, with AST standing for the Administrator for Space Transportation. It's a small organization that is mainly concerned with licensing commercial launch vehicles. It does some research into safety issues, but it is mainly a regulatory agency.

Ordinarily a government regulatory agency would seek to enhance its power by piling one rule on another. Battalions of specialized lawyers and lobbyists emerge to "protect" the industry from overregulation and all (Washingtonians) concerned are more or less happy. The FAA/AST is an exception, one of the few in the federal government.

This restraint was largely the work of one extraordinary African-American woman, Patricia "Patti" Grace Smith, who saw early on that overregulation would kill the emerging space industry. She made sure that the rules were written in such a way as to protect the public from errant rockets and debris, but that the companies and their customers could take informed risks without being squashed by regulations and predatory lawyers. Her first big success was in the space tourism area with her formula—which could be summarized as "kill yourself if you want to, but don't hurt the public."

She served during the Clinton and George W. Bush administrations and kept her personal politics quiet. It turned out she was a Democrat, but that hardly mattered at the time. The principles she embedded in her organization might have been written by a team of Cato Institute interns with input from Newt Gingrich.

The Space Guard's role, writing and enforcing commercial space safety regulations for things like asteroid mining or space hotels in low Earth orbit, will require specialists and also close cooperation with civilian institutions. The question of who will have the authority to deal with dangerous space junk is currently being fought over by the FAA/AST and the Commerce Department's Office of Space Commerce. Whichever entity wins, it most likely will eventually have to hand over that power to the Space Guard. Commerce won the fight for control of space traffic management, but that's probably a Pyrrhic victory.

The Coast Guard has a class of ships called buoy tenders which play a vital role in trying to keep ships from running into each other in crowed coastal waters. This maritime traffic management mission is analogous to the orbital traffic management mission. The Space Guard will surely have an important role trying to keep spacecraft and debris from colliding. To accomplish this mission, it will have to build relationships with other spacefaring nations, international institutions, and with a constantly growing and evolving set of private space operators.

Naturally, people will object to a military organization having this kind of power. Yet the Army Corps of Engineers has long played an essential role in managing America's inland waterways and in flood control. The only objections to the Army's authority are that it doesn't do enough flood control or spend enough on levees. Once the Space Guard is well established, it should be as uncontroversial as the Corps of Engineers.

A few years ago, it was suggested that a Space Guard be set up as a substitute for the Space Force. It was assumed this would be less provocative and would not push China and Russia into a space arms race. Since then, China and Russia have shown that no matter what the US does or does not do, they are determined, each for their own reasons, to build a variety of ASAT weapons. The need for a Space Guard as a separate but subordinate part of the Space Force cannot depend on what America's potential foes say or do. The role of the Space Guard as a quasi-military, quasi-civilian service is a function of America's broad national interests in space and not just diplomatic or commercial ones.

In the commercial field, as long as the Space Guard leadership sticks to the template created by Patti Grace Smith, it will do well. If they succumb to the kind of hostility toward industrial civilization and economic success we see at the Environmental Protection Agency, then the Space Guard will fail and probably end up embroiled in political polarization. This kind of ill will that we see so much of in Washington will surely paralyze both it and the industry and the settlements it is supposed to nurture.

We are slowly moving beyond the traditional space industry, whose most profitable activities involved sending various radio signals back and forth. The Space Guard role in providing safety support will be one of the best enabling mechanisms the US government could provide. NASA may have a role in technology development and in building habitable research bases on the Moon and elsewhere, but the Space Guard will have to create an inspection system that will include robotic and human safety advisers.

This system should aim at supporting the new economy rather than simply regarding it as a source of new tax revenue. It may take decades before space industries and settlements become profitable, but when they do, the cash flow will be enormous—but only if the efforts are not strangled in the crib. A Space Guard inspired by a mindset hostile to the whole idea of space colonization would be a disaster. If some of the anti-space ideologues in academia and the environmental movement are allowed to seize control of the institution, this disaster could be all too real.

This is particularly true in the field of space law. Left-of-center non-governmental organizations (NGOs) such as the Secure World Foundation are already working to undermine Trump's Artemis Accords. At the same time, China and the EU are both working in their own ways to impose a new set of space norms and so-called "best practices," which they hope will eventually be established as international customary law.

The attempt by the EU and the Obama administration to impose a space code of conduct—which would have inhibited America's ability to conduct military space operations while doing nothing to prevent those by the Chinese and Russians—is an example of what we can expect in the future. The Space Guard should, as an institution, be at least as ferocious a protector of America's commercial space interests as the Coast Guard is of America's fish.

The Coast Guard plays a large role in enforcing the laws (American and international) against maritime pollution. When we finally get around to deciding on an acceptable definition of what constitutes pollution in space and on celestial bodies, the Space Guard will naturally have a role in making sure that US entities conform with these laws and regulations.

However, it will be many years before the nations of the world come up with an acceptable set of definitions and rules. Spacefaring nations have already agreed on procedures to minimize and mitigate debris from launches and from civil and commercial activities. These practices are imperfect, and they do not cover military actions such as live ASAT tests.

For decades now the Air Force has been tracking spacecraft and debris in orbit, using radar and the Ground-Based Electro-Optical Deep Space Surveillance (GEODSS) system. Headquartered at Peterson AFB (soon to be Peterson SFB) in Colorado, its sensors are based at Socorro, New Mexico; Haleakalā Crater, Maui, Hawaii; and Diego Garcia in the Indian Ocean. The system performs a vital military function and thus cannot be completely turned over to the Space Guard, but the Guard should have a role in its operation and in the dissemination of the data derived from it.

No one should be under any illusions that international talks to control space pollution will be easy. One contentious issue is who should take part. Non-spacefaring nations will, in all likelihood, demand a seat at the table. Some spacefaring nations such as China may imagine that these "underdeveloped" nations will form a caucus that will hobble US and allied efforts to encode a practical set of rules. The idea would be that while the West is distracted by the negotiations, China, and perhaps Russia, can establish facts on the ground—the ground being the Moon's south pole—before anyone else.

Leaving aside for a moment the ruthless power politics involved in writing a binding set of pollution-control rules for space activities, the possibility of "space parks" should be considered. Some regions of the Moon, for example, should be made off-limits to all but the most minimal forms of human activity. The Apollo 11 landing site in the Sea of Tranquility is a good example of the sort of place that should be protected. And yet the whole idea of space parks is wide open to abuse by environmentalists who'd like to shut down as many human actions in the solar system as possible. Their slogan "There is no Planet B!" may be true right now. We should expect that the Greens will strongly protest any efforts to change that situation.

Handing over control of space traffic management to the Office of Space Commerce at the US Department of Commerce will, in short order, prove to be a mistake. The best one can say about the space people at Commerce is that they are all very nice individuals doing a limited job very nicely. One dreads to imagine them going up against the skilled international street brawlers who run the space-regulation elements inside the EU, not to mention the Chinese and Russians. These nice, inoffensive American civil servants would be crushed and, with them, any hope of regulating space commerce in US interests.

On the other hand, imagine a team of US Space Guardsmen and Guardswomen walking into a meeting in Vienna wearing their full uniforms and ready to fight like banshees to protect American space commerce from foreign attempts to regulate it out of existence (or at least out of profitability). We can expect the "international community" and its allies in DC will do everything they can to neutralize the Space Guard before it can be formed. The Space Guard should earn for itself a reputation for protecting America's commercial space interests with at least as much passion as the Coast Guard has for protecting America's fish.

Another possible role for the Space Guard is to support measures to protect American spacecraft, space bases, and people from the effects of solar and cosmic radiation. Over the last twenty years, NASA has done an outstanding job studying heliophysics, and there is every reason to believe that programs such as NASA's Living With a Star and others will provide a flow of knowledge that should inform all US space operations and help our international partners as well. Sadly, the same cannot be said of NOAA, whose inability to build new weather satellites or to integrate the space weather mission into its traditional duties has been obvious and unfortunate.

The Space Guard's role might include certifying radiation protection measures for manned spacecraft, space stations, and hotels and habitats on the Moon, Mars, and elsewhere. As the space economy grows and evolves, insurance companies will seek ways to reduce their risks, and having a US government agency with the expertise to decide what is and is not safe will make life easier for conventional investors.

This means that the Space Guard will have to have its own research facilities, as well as knit close relationships with major laboratories. A good model for the first is the Air Force Research Lab, and for the second, the Coast Guard's relationship with the Woods Hole Oceanographic Institute on Cape Cod. The need to ensure and certify that radiation protection coatings, devices, and methods actually perform as advertised is a matter of life and death for long-distance space travelers. Insurance companies will probably demand proof that commercial spacecraft and off-world facilities have state-of-the-art radiation protection before they agree to write policies. Having the world's best radiation protection certification service would be give the US a solid long-term advantage, similar to the one it had with the Food and Drug Administration pre-COVID-19.

If it does turn out that space solar power is economically feasible (and we should applaud the Space Force for supporting some modest experiments in that area), the Space Guard is the obvious organization to regulate the American firms and utilities involved in this business.

Clean solar power beamed from space to Earth, either by microwaves or lasers, has been a dream of space enthusiasts since Peter Glaser, a Czech refugee who ended up at Harvard, came up with the concept in the late 1960s. Gerald O'Neill popularized the idea as part of his High Frontier space colonization proposal (not to be confused with the High Frontier missile defense NGO) in the early '70s. At regular intervals the idea has been revived, but the obstacles are formidable indeed. Even the so-called "climate crisis" has not been enough, until now, to convince most governments, with the exception of China's, to fund serious research into this technology.

In the US, NASA seems institutionally allergic to even studying the idea. One cannot be 100 percent sure, but since NASA's budget is already strained by having to do too much with too little, the agency's leaders are appalled by the possibility of being handed another mission with little or no additional funding.

Mainstream environmentalists also line up against the idea; in the minds of the baby boomer Green leadership, it involves rockets and spacecraft and guys with short haircuts and skinny ties landing other guys with bad haircuts and flags on the Moon. For many Green political leaders, anything that gives a win to such people is a cultural and economic nightmare.

One role the Space Guard might play in the space solar power domain is to ensure and certify that the beams, either microwaves sent to receiver antennas (rectennas) or laser beams, if those turn out to be better, pose no danger to the population. Of course, some people will never be satisfied with any level of safety since they want the whole system to cease to exist, but if the Space Guard can certify that the beams are harmless or very nearly so, then the industry will find it possible to buy essential liability insurance. If not, then either the government will have to assume liability as it does (sort of) with the nuclear power industry, or the human race will never benefit from this clean energy technology.

Most designs for solar power satellites are huge. They may require either a single launch from a super-heavy launch vehicle or many launches for parts that would be assembled in orbit. In any case, they would be vulnerable to collisions with space junk. Here the Space Guard's role in regulating and, if necessary, removing debris would be critical. The Guard might not need to do the job itself; it could hire private contractors to carry out the mission, and this would eventually be paid for by taxes or fees on the space economy.

Another job will be to facilitate the transition of human-occupied commercial facilities from business ventures to self-governing communities. The successful way that Massachusetts switched over from a seventeenth-century English company with a royal charter to a colony with its own quasi-legislature and governor could be a model for the mid-twenty-first century.

It is extremely unlikely that any member of the USSF will actually go into space in order to perform a strictly military mission for many, many decades to come. If all goes well, Space Guard inspectors, safety personnel, law enforcement, and (say it softly) tax collectors will probably become regular visitors to American space stations and settlements sometime in the 2030s or '40s. Obviously, Space Force personnel will be eligible to apply for NASA astronaut positions just like the members of the other services can. In spite of advanced communications technology, video conferencing, and blockchain accounting procedures, human presence will be needed wherever humans are involved, if only for political reasons.

For the Space Guard, as for the Space Force, the decision on how and when to incorporate a reserve component is going to have a big political impact as well as an operational one. A reserve component may have a disproportionate regional aspect. There will be lots more Space Guard reservists in places like Florida and California than in places like New York or Michigan. Yet the talent the Space Guard Reserve needs will not necessarily be concentrated in the places where the space industry is.

With any luck, the Space Guard Reserve will be an elite organization with a big contingent of experienced and qualified space law experts. These men and women could be called on for advice and, if necessary, to argue cases in US courts.

PART
THREE

OTHERS

CHAPTER 9

CHINA AND RUSSIA

PART 1: THE CHINESE CHALLENGE

In his highly insightful study *1587, A Year of No Significance: The Ming Dynasty in Decline*, Ray Huang wrote the following glimpse of classic Chinese government thinking: "The Empire was not set up to wage war, to reconstruct its own society, to conduct a national program of any kind, nor even to improve the standard of living of the populace aside from taking preventative measures against famine. Its purpose was to maintain peace and stability."[38]

Recent Chinese hi story shows deviations from this norm, particularly between the final collapse of the Manchu (Qing) Dynasty in 1911 and the death of Mao in 1976. After that great murderous revolutionary was dead, China reverted to its traditional "peace and stability" polices under Deng Xiaoping and his successors. However, under the current leadership of president Xi Jinping, China's government seems to be rejecting the old attitude that held "we are the center of the world, so why do we need any foreign stuff?"

Now it looks more and more like China is acting as if it were a rising power on the model of pre-1914 Germany. Its aggressive diplomacy and naval buildup all point in this direction. However, nothing better indicates China's seriousness about challenging the global *Pax Americana* than its military space program.

During the Mao era, there were the three "musts" that China had to have: an atomic bomb, a hydrogen bomb, and a satellite. In 1969, the People's Republic of China was one of only two nations (the other was Fidel Castro's Cuba) that ignored the US Apollo 11 Moon mission. In April 1970, China successfully used a modified IRBM to put a satellite into orbit.

By the time Nixon made his historic trip to Beijing in 1972, China was even more technologically backward compared to the West than it had been during the Ming Dynasty. The Cultural Revolution had destroyed a generation of scientists and engineers; China desperately needed access to US and Western technology.

They got it.

When US officials got a close look at the state of the Chinese military, they were amazed at how technologically backwards it was. Its most advanced systems were poor copies of late-1950s Soviet weaponry. Most of the huge People's Liberation Army (PLA) was equipped with technology that was barely out of the 1930s and '40s. Helping China update its forces so it could represent a real threat to the USSR, or at least resist a Red Army attack in Manchuria or Xinjiang, was a US priority.

Ronald Reagan said it clearly: "Our intention is to provide China with a capability to defend itself more effectively against the common threat to the region."[39]

Two things happened in the 1980s: One was the near total abolition of Maoist economic policies, and once the collectivist agricultural policies were gone, China's rural poor suddenly began to produce food in amounts previously undreamed of. This was partly due to China getting access to Western agricultural technology, but it was mostly caused by the absence of communism and its replacement by "socialism with Chinese characteristics," which looked a lot like mercantilist capitalism. The second thing was the across-the-board failure of Soviet communism and the collapse of the threat on China's northern border. Together these two events liberated resources that could go into "development" and the accelerated modernization of the PLA.

By 1989, China had gotten into the business of launching commercial satellites and was competing with the European Ariane rocket and with Russia. Then came Tiananmen.

The sanctions that were reluctantly put on China's military and space industry by the George H. W. Bush administration and its allies put an end to the quasi-alliance that had existed between the two powers. The commercial relationship, however, thrived. By 2010, China was the "workshop of the world." It seemed as if its economy would continue to grow at 10 percent per year into the indefinite future.

After Xi Jinping became head of the party and the state in 2012, Chinese strategy moved away from the Deng-era idea of "peaceful rising," which assumed that they would rise to take their place as one major power among several in a vaguely liberal world order, to a new goal—the China Dream. As graduate student Christopher Stone put it, "This China Dream is the re-establishment of the Sino-Centric order, or in other words, harmony in place of the friction that dominates the United States-led international system under the UN Charter." Before Xi consolidated his power, some Chinese military officers were heard to say that they regarded the Chinese Communist Party as just another dynasty in China's long history, with the implication that there would be other dynasties after the communists. For others, the very idea that one day the communists would be overthrown was outrageous. The implications of this split inside the PLA have not, to my knowledge, been properly assessed.

China's space industry, in all its aspects, seemed to be thriving. In 2003, it launched its first "taikonaut" into orbit. Since then, the Chinese have launched and operated a pair of basic, Salyut-style space stations and are planning larger ones—and eventually manned missions to the Moon and Mars. China's space exploration program is, it must be emphasized, no more aggressive or hostile than America's or, for that matter, anyone else's. The race to explore and exploit the resources of the solar system does not imply an imperial-style competition. China, like any other spacefaring nation, intends to have its say on the issue of property rights in space. For the foreseeable future this is an issue that will be left to diplomats and businesspeople, not to the military. At least for now.

In the military sphere, China seems aggressive and determined to use space to undermine US and allied military effectiveness and to empower its own forces. The Chinese see America's reluctance to actively defend its own space assets and to build up its own set of space weapons (aside from a minor "counterspace device" jamming system) as a weakness to be exploited. One very experienced military space expert described "decades of jamming, lasing, and other direct-ed-energy attacks against satellites or the signals they carry."[40] The vast majority of those attacks has been credibly attributed to China. For those without access to classified information, it is an open question as to how much damage has actually been done to US space assets. On the one hand, we don't see much evidence that US military space operations have been impeded, so it is possible that China is carrying out these attacks simply to annoy the US and remind Washington that China must be taken seriously. On the other hand, it could be that US systems have been degraded and their operational lifespans shortened by these Chinese actions. If the latter is true, then China will have cost the US military billions.

China's notorious ASAT test in January 2007 not only showed off its weapons capability, but also created more space debris than any other single event since Sputnik. Since then, Chinese weapons tests have not included blowing up any physical objects, but their simu-lated tests and what appear to be "proximity operations" prove that, in spite of the Obama administration's attempt to convince them to show restraint, the Chinese space weapons program has not slowed down. The Chinese leadership seems to believe that the advantages of having a full array of space weapons is worth more to them than any level of "strategic restraint."

China's approach to war in space is relentlessly practical—or at least it seems that way at first glance. The Chinese see America's use of a wide variety of space systems as creating an information vulnerability. The twenty-first-century US way of war seems to them to depend on an unimpeded rapid flow of data between all US military elements. The Chinese believe that if they can interrupt or degrade this flow, they will be able to isolate and defeat the separate parts of US strength. The US military is perfectly aware of this chink in its armor and has been working to fix it for more than twenty years. America's politicians, especially during the Obama administration, have tried to avoid dealing directly with the problem, but the military has tried to work around the obstacles thrown up by the White House.

Meanwhile, China, like any self-respecting twenty-first-century power, has been building up its own array of on-orbit assets. In October 2003, China joined the exclusive club of nations that had sent people into orbit and earned the right, according to terminological convention, to have its own designation for the people it sends into space; in China's case, they are called taikonauts. Yang Liwei's flight in a Shenzhou capsule launched by a Long March 2F rocket was a technical, economic, and political triumph for the regime. While the technology was not 100 percent Chinese (Shenzhou owed a lot to Russian help), it marked the beginning of China's drive to build a Chinese presence in the solar system. They followed up with a small space station and will soon try and build a larger one, and they plan to have a manned base on or near the Moon's south pole sometime within this decade.

Their robotic mission to the dark side of the Moon showed considerable sophistication, particularly their use of a communications relay satellites based at the L2 Lagrange point beyond the Moon's orbit. China's BeiDou PNT constellation certainly matches Russia's GLONASS, if not the US GPS system, and it has built up a comprehensive array of civil (Gaofen—as part of the China High-Resolution Earth Observation System, or CHEOS) and military (Yogan) Earth observation and sensing satellites. Their plan to have the UN install them in the UN Global Geospatial Knowledge and Innovation Center in Zhejiang Province—along with a UN-supported "big data" center—

shows just how determined the Chinese are to play a large, perhaps dominant, role in the field of Earth sensing science and technology. And, of course, the intelligence advantages to having these meters in China cannot be underestimated.

The PLA has created a Strategic Support Force, which includes its military space forces. This force is focused on all aspects of information warfare, including cyberwar, influence, and deception operations. It looked at first like an impressive concept, putting everything to do with information into a single command, and it may indeed prove to be a stroke of genius on the part of the PLA leadership. But it may also weaken the cyber defenses of the other parts of the Chinese military, depriving the air force and navy of their own cybersecurity experts and preventing China's military space forces from focusing on operations and technology that are not directly connected to "information operations."

The increasing dependence on civil and military space assets creates a dilemma for China's leaders. Against the United States, their ASATs could, in theory, cripple its military power. Against any other power, particularly India, China's need to use its space assets to achieve military superiority creates a real and growing vulnerability to other nations' space weapons. Put another way, China cannot be a great global power without its own comprehensive set of space assets. At the same time, it cannot fight a war against any power other than the US where those space assets are not liable to be attacked in ways that would degrade China's military strength more than it would the strength of its potential adversaries.

As of 2020, the PLA looks very impressive indeed; it seeks to dominate East Asia in ways no power, except the US, has since the age of Vasco da Gama. If China's economy continues to grow the way it has done, it has an excellent chance of becoming the undisputed regional hegemon. However, between Trump's economic policies, the slow awakening of a new and more skeptical attitude toward China's commercial policies in Europe and elsewhere, the increasing demands of the middle class for a better standard of living, and the effects of the COVID-19 virus, China's days as a rapidly rising power may be over. It will be a while—probably a decade—before this impacts China's space programs, but if China continues on the path it has been on recently, it will find the balance of power moving back in Washington's direction.

Yet as baseball sage Yogi Berra is supposed to have said, "It's tough to make predictions, especially about the future."

PART 2: RUSSIANS RISING

No nation on Earth, except possibly the US, has more of its national ego wrapped up in space activities than Russia. This is almost wholly due to Sputnik and to Yuri Gagarin.

After the 1917 revolution, Soviet Russia was forced by the new ruling class to embrace Marxism, a political philosophy which claimed to be scientific. This led to an interesting situation where the Communist Party was perfectly happy to support scientists who were "politically correct," but were also happy to clamp down hard on scientists, engineers, and researchers who failed to conform to the party's ever-changing ideological line. The party was normally willing to tolerate minor deviancies in men who were working on new weapons, but even the best minds in the country were vulnerable to punishment if they dissented from the state's philosophy—or were even suspected of being slightly unorthodox.

Sergei Korolev's life story is a near-perfect illustration of how this worked. A brilliant engineer, Korolev was arrested by the secret police in June 1938 and sent to the arctic gulag camp at Kolyma. He barely survived. He later was imprisoned in a secret research center in Moscow and ultimately became the chief rocket designer for the Soviet Union. Korolev died in 1965 due, at least in part, to the brutal treatment he'd received as a prisoner. One could speculate that if he'd not been so mistreated, the USSR might have beaten the US to the Moon. On the other hand, it is also evident that if he had not been beaten and tortured, the USSR would not have been the USSR. In any case, Korolev has gone down in world history as the man who launched the first satellite into orbit.

It is still an open question whether Eisenhower deliberately let the Soviets launch the world's first satellite in 1957 because he wanted to establish the principal of freedom of space so that US spy satellites could fly freely and legally over the USSR. What is indisputable is that once the US began to regularly fly reconnaissance spacecraft over Russia, China, and other totalitarian states, they lost much of the advantage they had long held by being closed societies.

However, it is evident that Ike was not prepared for the gigantic propaganda victory that Sputnik handed to Moscow. Indeed, Khrushchev and his fellow politburo members could not believe their luck. Not only did the Soviets successfully use their well-oiled propaganda machinery to tout their success and project an image of themselves as a scientific and technologically advanced society, but in the US, Eisenhower's image was effectively tarnished for the rest of his term by the Democrats and their allies. Lyndon Johnson, who was then the majority leader in the Senate, jumped onto the space bandwagon.

The idea was propagated that the US was a rich, lazy consumer society which neglected its military and the scientific education of its youth. Rock 'n' roll, no rockets, and car tail fins (not missiles with tail fins) was supposedly the image that America presented to itself and to the rest of the world. The men in the Kremlin could not have asked for better proof that the future would be Soviet. As Khrushchev put it, "We will bury you."

Meanwhile, in the hard world of military power, Russia soon discovered that building a force of ICBMs was harder than it looked. The R-7 rocket proved unreliable and very expensive, but as a space launch vehicle it proved surprisingly effective, and today's Soyuz rocket is both fairly reliable and robust. The accident which caused a Soyuz capsule to abort its mission to the International Space Station in October 2018 proved that the Russian escape system for its spacecraft worked.

The importance the Kremlin put on ICBMs was explained by Khrushchev in his memoirs: "Our potential enemy—our principal, our most powerful, our most dangerous enemy—was so far away from us that we couldn't have reached him with our Air Force. Only by building up a nuclear missile force could we keep the enemy from unleashing war against us."[41] Until roughly the mid-1960s, the Soviet ICBM force was mostly a bluff; this should have been obvious when the Russians chose to move most of their medium- and intermediate-range missiles to Cuba in 1962. If they had had a fully operational set of ICBMs, they would never have needed to make the risky and provocative move into Cuba.

Beginning in 1958, as a response to Sputnik, the US began to massively invest in ICBMs, space launch vehicles, and satellites. By the early 1960s, the US had effectively established its technological superiority, but thanks to America's longstanding inferiority in propaganda and political warfare, it was consistently unable to translate this advantage into usable leverage in the Cold War—until Ronald Reagan made his famous "Star Wars" speech in March 1983.

In any case, the military needs of the Soviet armed forces were not neglected. In 1962, they launched their first reconnaissance satellite using Soyuz capsule technology and following the US example of dropping the exposed film back to Earth to be recovered and developed. Later, they perfected electronic intelligence-gathering satellites and built a nuclear-powered ocean reconnaissance satellite system that involved two satellites, launched within days of each other, that both carried powerful radars which could locate ships in any weather conditions. This system came to public attention in January 1978 after Cosmos 954 became uncontrollable and the satellite and its reactor crashed in northern Canada, causing considerable international concern. However, that concern was quickly forgotten, a tribute to the value of the communist empire's "soft power."

Until 1983 the USSR believed that it was either superior to the US in spacepower or that America was only slightly better off. The US may have had better reconnaissance satellites, but the Soviets had a monopoly on ASATs. In order to fulfill their strategic objectives, the Soviet armed forces did not need much in the way of space-based communications or navigation. In spite of several glitches, their early warning satellites were, they believed, satisfactory. This, combined with their overall superiority in nuclear and conventional firepower, convinced them that the triumph of international communism was only a matter of time. In the mid-1990s, Air Force General Thomas Moorman explained that while the US failed to develop a consistent ASAT development program, the Soviets, "developed a primitive but highly visible, co-orbital 'killer satellite' interceptor that could be used to attack and destroy space satellites in low Earth orbit. From 1968 to 1982 the Soviets achieved a 50 percent success rate with these anti-satellites."

Then came "Star Wars." In March 1983, president Ronald Reagan gave a speech that was mostly devoted to asking Congress to fund the MX heavy ICBM but ended with a dramatic plea to America's scientists to find a way to make nuclear weapons "impotent and obsolete." Reagan was, by all accounts, perfectly sincere, but the Soviets believed this was a signal to the world that the US was going to try and quickly develop space-based anti-missile weapons. At the time, Reagan knew the Soviets had an extensive ASAT program; in 1985, the US administration wrote, "The USSR has had for more than a dozen years the world's only operational antisatellite system, a co-orbital device which enters into the same orbit as its target satellite and, when it gets close enough, destroys the satellite by exploding a conventional warhead."[42]

The Soviets were obsessed with America's technological prowess, and at great expense they tried to match it—or at least create the illusion that they were matching it. Their attempt to build the Buran space shuttle, for example, was a gigantic waste that only proved to the world how incompetent they were. Even worse was the psychological impact caused by the need for Moscow's diplomats and spies to haunt Western toy stores in order to buy toys which could be taken apart so that the Soviet military industry could salvage the microchips embedded in things like Pong or Space Invaders. Imagine for a moment how the leadership in the Kremlin must have reacted when they were told that the toys and games available to children in Chicago or London or Lyons had superior information-processing technology than did the command systems on their most impressive warships or in the headquarters of their frontline tank divisions or on their most sophisticated satellites.

However, when they concentrated on making incremental improvements to existing hardware, they were on solid ground, and today much of their best space technology is derived from items they created back in their glory days of their space program. The Soyuz rockets that are launched from the French base at Kourou are, in effect, highly developed versions of the R-7 ICBM which was first tested in 1957. Russia's GLONASS PNT satellites are in every way inferior to the GPS system, but, like GPS, they are essentially a military asset and Russia has no reason to abandon the program.

A case could be made that Russia is even more dependent on space systems for its national security than the US. Its conventional armed forces are a fraction of the size they were during the Cold War, and its tank forces are not only much smaller, but Russia has only been able to build a tiny number of its most advanced tanks. The bulk of its armor is made up of T-72 tanks and their derivatives. Russia simply does not have the forces to threaten NATOs Eastern and Central European members, dominate the Caucasus and Central Asia, and deter China in the Far East. This conventional weakness can only be compensated for by a variety of nuclear weapons and delivery systems. These, in turn, rely on space systems for their effectiveness.

Yet Russia's space industry is in long-term decline. We got a demonstration of this in October 2018 when a Soyuz capsule launched from the Baikonur base in Kazakhstan failed to reach orbit. Fortunately, the crew landed safely, but such a failure of what had previously been a reliable and robust system was a sign if how bad things have gotten. One presumption is that the Soviet-era scientists, engineers, and technicians, who once made up a privileged elite, have retired and not been replaced, or that their successors are just not as good. Since the Russian space industry can no longer recruit from places like Ukraine and the Baltic states, not to mention the rest of the Soviet empire, the pool of high-quality employees is smaller than it used to be.

Lower relative pay and less prestige have combined to put unbearable pressure on Russia's space industry to find ways to perform in all the domains of spacepower. The failures are not surprising; what is surprising is that there are not more of them. Obviously, Russian president Vladimir Putin puts a high priority on space activities. For example, he refused to allow the GLONASS PNT system to fade away and instead put considerable resources into rebuilding the satellite constellation.

CHAPTER 10

ASYMMETRIC AND LOW-INTENSITY WARFARE IN SPACE

During the Vietnam War, a story was told about a Green Beret officer who'd been invited to the Air War College to speak on the subject of "Airpower and Counterinsurgency." He is supposed to have begun by explaining to his audience that "the best use for a supersonic fighter-bomber in a guerrilla war is to drag it into a village and hold a competition among the village women to see how fast they can turn the aluminum into pots and pans." He recognized that, in war, technological superiority cannot substitute for political will. However, by the 1960s, airpower was so integral to almost all forms of military activity that it was impossible for any conventional army or navy to fight any sort of campaign without it.

The creation of a United States Space Force in 2019 has done little to answer the question of how space technology can be used specifically to help win the kind of low-intensity conflicts that are endemic to the twenty-first century. It is perfectly natural for the USSF to want to concentrate its efforts on dealing with the challenges from Russia and, above all, China, but in spite of the desire of America's political leadership to avoid or withdraw from these dirty little wars, we are going to find ourselves involved in them for a long time to come.

In the introduction to the 1983 English-language edition of Giulio Douhet's 1921 *The Command of the Air*, the editor wrote "In that

[Douhet's] theory, airpower became the use of space off the surface of the earth to decide war on the surface of the earth." According to Douhet, the way that aviation had been used during World War I was "illogical." The extreme case he made for strategic bombing—and nothing else—as the key to victory in future wars was based, he believed, on "sound and orderly" logic. Historically, we know that nothing in human affairs is less logical or orderly than war. Technological opportunities may be grasped or not, depending on policies that are made by fallible human beings.

Douhet got a lot of things wrong, but one important thing that he got right was that aviation would become a "great industry" and that "it constitutes a means of political power, national wealth, and military security." A century later, aviation is all this and more; it helps drive development in sectors such as information technology and exotic matériel, to name only two. Without it, the human and political geography of our planet would be very different from what they are now.

The debate on whether airpower or seapower is a better analog for spacepower is ongoing. Spacepower is a new and little-understood element of national power. For America, future decisions about military space systems and doctrine will be shaped, in large part, by the challenges that present and future foes pose to the complex, expensive, and ever more valuable amalgam of people, hardware, and ideas we have come to call "spacepower."

In fact, all three (seapower, airpower, and spacepower) have commercial, scientific, colonial, sporting (leisure), and military aspects. No nation can be said to have true seapower without a merchant marine, oceanographers, marine biologists, yachtsmen and -women, bases and overseas assets, and a variety of fighting ships, aircraft, and amphibious forces. Airpower relies on commercial aviation, numerous scientific disciplines, sports fliers, air bases and airports, international agreements on access and safety, and satellites for communication, reconnaissance, navigation, and other purposes.

Technology developed for non-military space purposes will probably play an even greater part in future military space systems than it does today. One possible example is suborbital space tourism vehicles, such as Burt Rutan's SpaceShipTwo (currently in development for Virgin Galactic), which could be test beds for technologies that will be used on a long-range hypersonic bomber or USMC SUSTAIN-type transport circa 2035.

Command of space, in the grand strategic sense, is said to be achieved when the enemy can no longer hamper one's use of celestial lines of communication and is unable to defend his own. With few exceptions, general command enables the unlimited use of space for diplomacy, commerce, military operations, and related reasons.

Today, there can be no question that spacepower is as important, in its own way, to the American way of war (and civilian way of life) as airpower was in the 1960s. While the US has not yet moved to develop or deploy systems that can be termed "space weapons," it is coming awfully close. Sometime soon it may decide to build both defensive and offensive counter-space weapons as well as space-based missile defense systems—and possibly even direct space-to-Earth weapons. Whether the US does or does not choose to "weaponize" space, it will not be the first nation to do so. The USSR was the first nation to put a weapon into orbit when, in 1974, it launched a Salyut space station equipped with a 23 mm automatic cannon.

Since the attack on September 11, 2001, the US and its allies have been engaged in a complex, new kind of war. It has been called the Global War on Terror or the Long War; in any case, the old doctrines and theories of spacepower have got to be reexamined in light of our nearly twenty years of experience. It is simply impossible to imagine the US—or any other advanced nation—fighting a war without using space to one degree or another. America's enemies have moved quickly to use what space assets they can, from GPS to satellite telephones and television. As far as we can tell from unclassified sources, there has been little if any attempt to deny them the use, for example, of satellite television. As time goes on, this may change. In any case, it can be confidently anticipated that there will be future requirements for systems that can selectively jam or incapacitate ostensibly civilian communications and direct-broadcast satellites.

The nature of spacepower is such that, even more than with other forms of warfare, the technological and the political decisions are intertwined. A good example of a politically driven spacepower project is the European Galileo satellite navigation system, which was developed as a response to the success of America's GPS system during the 1991 Gulf War. The technology was available and was well within the capability of Europe's space industry. The desire to spite the US after 9/11 was strong enough to overcome the objections of some of America's traditional European friends. To paraphrase what one British commentator wrote at the time, "Osama has helped Europe win the space race." The principal lesson for the US is that spacepower—in both its military and civil aspects—has, in the past, derived more from *political* judgments than from technological or military ones.

This was certainly evident in the Eisenhower administration's attitude toward building the first satellite: "[P]lacing a satellite in orbit amounted to a statement about the ability to deliver nuclear weapons over intercontinental distances. Also, no one knew whether satellites would be considered violations of extended sovereign airspace. Unprecedented policy issues needed to be resolved for international acceptance of satellites."

The dynamics of the future international situation will, alongside internal US political judgments, determine if, when, and how the US chooses to develop its spacepower and future space weapons. The War on Terror and other possible conflicts will drive short- and medium-term requirements. The foes we face today have shown that they are determined and imaginative. Entities such as al-Qaeda and Hezbollah and their offshoots, imitators, allies, and sponsors can be expected to do whatever they can to harm the US and, above all, its military power. America may be fighting a limited war with well-defined rules of engagement, but for the enemy, it is a total and unlimited war.

Whatever happens in the US, other nations and entities will build, deploy, and use space weapons according to their own priorities, time-tables, and perceived self-interest. Spacepower is increasingly seen as an essential part of a nation's warfighting capability, and all but the poorest countries aspire to have at least a toehold in orbit. According to a British academic, military spacepower theory during the Cold War was divided into four schools: the sanctuary school, the survivability school, the space control school, and the high-ground school.[43]

The sanctuary school believed that space should not be weaponized. This school believed that the value of space assets for arms control and strategic stability is such "that space must be kept free from weapons, and antisatellite weapons must be prohibited, since they would threaten the space systems providing these capabilities."

The survivability school was based on the idea that space systems "are inherently less survivable than terrestrial forces." This school believed, however, that space forces "must not be depended on for those functions in wartime, because they would not survive." In the context of a possible US-Soviet nuclear war, this made a certain degree of sense, even though both sides made efforts to harden some of their critical systems.

The space control school used the analogy to seapower to argue for the development and maintenance of spacepower. Starting from Alfred Thayer Mahan's analysis of eighteenth- and nineteenth-century British seapower, the space control school chose to "argue that there are space lanes of communications like sea lanes of communication that must be controlled if a war is to be won in the terrestrial theaters…. In future wars space control will be coequal with air and sea control."

The high ground school supposedly was based on the old military axiom that "domination of the high ground ensures domination of lower lying areas." Perhaps it would be better to call this the "gravity school" since it emphasized the primordial importance of simple

Newtonian physics. In space, as on Earth, (kinetic) striking power, mobility, and defensive capability all depend, to various degrees, on gravity for their effectiveness or lack thereof.

In 2001, subsequent to the Rumsfeld Commission report on the management of US military space, the Air Force was named the "Executive Agent for Space." This move hardly changed anything. The Air Force remained fixated on its old airpower doctrines.

Space was still the "New High Ground." Programs such as SBIRS and AEHF remained troubled, the Navy continued to develop its own communications satellites, and the NRO and National Imagery and Mapping Agency (now the NGA) continued to run the imagery and intelligence collection satellites. As "executive agent," the Air Force found it no easier to convince Congress to fund the space-based radar (SBR) or to look kindly on the Transformational Communications Satellite (TSAT).

There is an emerging consensus that the main cause for program failure (or near failure) is an out-of-control requirements process and a broken cost estimation system. Fixing these problems will not only help America's space forces, but it will solve problems that plague the air, land, sea, and cyber components of American national power. The difficulty is that while incremental improvements to existing systems and technologies are a safe and relatively low-cost way to proceed, revolutionary developments are more likely to produce significant impacts on the battlefield. For example, GPS was a leap ahead of any existing navigation system when it was proposed in the 1970s. It is a prime example of a revolutionary system that could not have been produced, in its actual form, by a step-by-step development process.

What has become clear over the past few years is that the co-called "sanctuary consensus" against the "weaponization of space," which some scholars assumed existed during the middle years of the Cold War, is gone. It has been replaced by a vigorous debate which encompasses all aspects of the issue, including defensive and offensive counter-space systems, space-based missile defense weapons and space-to-

Earth weapons. The only agreement that still enjoys near-universal support is the part of the Outer Space Treaty of 1967 which prohibits the stationing of weapons of mass destruction in space.

The October 2006 National Space Policy put out by the Bush White House threw cold water on the whole idea of diplomacy and of new international treaties to control space activities. "The United States will oppose the development of new legal regimes or other restrictions that seek to prohibit or limit US access to or use of space. Proposed arms control agreements or restrictions must not impair the rights of the United States to conduct research, development, testing, and operations or other activities in space for US national interests...."

America's military space systems have not been the decisive force in what used to be called the Global War on Terror that some had hoped they would be. Yet without them it is impossible to imagine the US military being able to fight anywhere near as effectively as it has. Thanks to its use by JDAMs, GPS has produced a revolution in low-cost precision weaponry; communications satellites make long-range UAVs such as the Predator, Reaper, and Global Hawk into genuinely revolutionary assets. The enemy's style of low-intensity, asymmetric, terrorist warfare has meant that the military space hardware developed for conventional (and nuclear) warfare and their associated Earth-based organizations and support networks are not as useful as they should be, particularly given the costly investments that have been made and continue to be made in these systems.

One defense expert, Loren Thompson of the Lexington Institute, said that "terrorists may present a target too fleeting to follow with satellites," and that he sometimes wonders whether some of the funding planned for next-generation imagery satellites "would be better utilized if it was diverted to...human intelligence." Aside from the legal and cultural issues that tend to discourage American human intelligence efforts, this judgment may be as premature as the ones that have, for many decades, predicted the demise of the tank. The value of view from the high ground is not going to change; what will inevitably change is the way that information is exploited and distributed.

The dual nature of asymmetric—or, as the Iranians call it, unbalanced—warfare is that it is tactically based on extreme cruelty; for example, killing whole families in the most horrible way possible to warn people not to cooperate with the authorities or the Americans. Mohamed Farrah Aidid, the Somali warlord whose attempted capture in 1993 led to the "Black Hawk Down" incident, is quoted as explaining the methods of his hero, the "Mad Mullah," thusly: "Any person or tribe who did not agree to submit to him would be declared *kufr* [irreligious] and he would unhesitatingly order his Dervish soldiers to kill them and their children and women and snatch all their property."

This is combined with a strategy that is aimed directly at the enemy's political decision makers, the use of a friendly or coerced media to misinform, and the use of the so-called "cultural high ground" to produce artifacts such as movies, art, documentaries, books, comics, etc. that support the aims of the asymmetric warriors. The reluctance of the US to deploy the nonlethal, high-powered microwave weapon known as the Active Denial System is an example of the effectiveness of this media-cultural strategy.

The limits of airpower in low-intensity conflict or asymmetric warfare are well known. In 1920, the British tried to use their World War I air force against the tribes in the Afghan-Indian (now Pakistani) border region. Air Commodore Tom Webb-Bowen, who commanded the Royal Air Force during the campaign, stated that the "RAF acting alone will never overcome a courageous people." Despite this, attempts are constantly made, as Israel did in July 2006 in southern Lebanon, to use airpower and airpower alone against asymmetric foes.

For good moral and constitutional reasons, the US military does not normally fight its wars with the direct aim of inflicting heavy civilian casualties on its enemies. Historically it has failed to fight political wars in a politically effective fashion. Modern warfighters need the counter-propaganda that is being supplied free of charge by the new alternative media in the blogosphere and elsewhere. The fact that the bloggers are not under any sort of government control makes them far more effective than would be the case if they had to clear their "work product" through the bureaucracy. Sadly. America's enemies are able to use the internet and social media for their own purposes.

Likewise, military planners should not expect that strategic communications involving space warfare decisions will be effective if controlled by the DoD. Advocates and adversaries of these decisions will take their cases to the public in a messy and unsupervised debate with few, if any, limits.

Interestingly, in the case of the Israel-Hezbollah War in July 2006, the blogs succeeded in undermining Hezbollah's propaganda in ways that neither Israel nor the US ever could. This is a rare example of Westerners practicing asymmetric information warfare against the Islamists and their allies in what is referred to as the mainstream media. Interestingly, the bloggers were able to use publicly available satellite imagery to disprove some claims about the effects of Israel's bombing campaign and to put the so-called "Qana massacre" of 1986 and the associated staged photos into context (see Richard North, EU Referendum blog).

The use of these pictures and others like them in uncoordinated and unconventional ways is only part of the change in the nature of spacepower brought about by the internet and the "new media." Years ago, spacepower was highly centralized, and many aspects of it were highly classified. Today, it is possible for anyone to go on the internet and find out when some US (or other) spy satellites are going to be overhead. Encrypted satellite communications are available and affordable; it may soon be possible to counter navigation warfare use of GPS signals, and the advantage the US military has long gained from its highly advanced weather satellites may also no longer be what it once was.

The historic analogy between seapower and spacepower is sometimes a more useful one than the comparison of airpower and spacepower. This is due, in part, to the time scales involved; a space mission may last for months or years or even decades, just as the voyages of the old sailing ships did. The weather and the tides and currents that an eighteenth- or nineteenth-century sailor had to contend with are closer to the orbital mechanics and radiation storms that a space operator must deal with than to those things that an air campaign planner has to take into account.

Our knowledge of the space environment is about as hazy today as the British Admiralty's understanding of climate and oceanography were two hundred years ago. We have only begun to understand such phenomena as the Moon's "weak stability boundary" or the full effects of solar coronal mass ejections (CMEs). NASA and NOAA's research on space weather and on the sun-Earth connection, done under the Living With a Star program, is a direct contribution to America's over-all spacepower.

In the case of space and the War on Terror, one way of thinking about it is to compare the terrorist strategy to the commerce raiding sometimes referred to as the *Guerre de Course*. In both cases, it is the weaker party that chooses to fight against essentially civilian targets due to its inability to effectively engage the armed forces of the enemy. The fact that one of the prime terrorist target sets has been civil aviation reminds one that the commerce raiders of old, whether pirates or privateers, aimed at attacking essential routes and methods of communication. Since so much commerce now moves electronically, either over fiber optic cable or through communications satellites, space commerce will be a logical target set for our enemies. Indeed, they may launch attacks that seem to us to be counterproductive.

For Germany in World War I, the decision to use unlimited submarine warfare in an attempt to strangle Britain might have worked, in spite of the fact that this caused America to enter the war. The Royal Navy's lack of adequate or reliable anti-submarine weapons and its reluctance to institute convoys, which were seen as an essentially defensive distraction from the Mahanian main battle, gave the U-boats their chance. In 1917, they nearly knocked Britain out of the war due to sheer starvation. The food-rationing instituted at the time was a drastic and panicky response to the situation.

In the future, the US may find that its lack of defensive space weapons and the reluctance of the Air Force or the political establishment to invest in space systems protection may lead to a similar situation, with no guarantees that the right decisions will be made in time to salvage the situation. In this case, the US will not suffer from lack of food, but from a shortage of communications capability and timely and reliable information.

Protecting space assets is not something that comes naturally to peacetime military planners. Despite 9/11 and the subsequent conflicts, US military space operations and procurement are still run on an essentially peacetime basis. The added expense and weight needed for shielding against EMP or for active self-defense systems are seemingly too hard to justify and pay for. Politics and budgets are part of the problem, but US military space doctrine and a reluctance to imagine things from the enemy's point of view may be equally to blame.

Since space and, particularly, military space assets provide the US and its allies with critical advantages over their enemies, especially in terms of the ability to shape and prepare the battlefield, it is to be expected that America's foes will try hard to eliminate this advantage. Hard-target attack methods such as co-orbital or direct-ascent anti-satellite weapons may not be beyond the ability of states such as Iran or North Korea, who can, after all, build the long-range missiles that could theoretically carry such weapons into space, but they would probably prefer to use less obvious and more deniable means of attack.

An enemy's goal will be to eliminate or degrade US space systems. If this can be done without kinetically attacking the spacecraft, then the foe will probably prefer to use covert means. Deniable jamming attacks, such as those used by Iran from a base in Cuba or by Israel during the Hezbollah War against commercial broadcasting satellites, are examples of this type of space warfare. Another example is the reported Chinese blinding laser attacks against US military satellites of which, according to sources, there have been several tests over the past several years. In 2019, as part of its new military space program, France announced that it would build lasers similar to the ones China has been using.

From our point of view, this limited form of conflict may be logical, but from a different political vantage point, it may be preferable to blatantly and violently knocking out one or more US satellites and taking the consequences. This would be, in the enemy's eyes and in the eyes of the Arab/Muslim/Third World masses, a humiliation for the US. Even if American retaliation were to be swift and painful, something that is by no means assured, an enemy might find it worthwhile in its own internal, regional, and religious context.

In the 1920s and '30s, Middle Eastern tribesmen fired millions of rounds of rifle ammunition at the vulnerable biplanes of the RAF. Occasionally they managed to hit a vital spot and brought one down; these "victories" had little propaganda effect due to the remoteness of the fighting and to the lack of modern media that could amplify the impact of a single loss. Today, after the Black Hawk Down battle in Mogadishu and its aftermath, tribesmen, terrorists, and rogue states are acutely aware of the ability of the world media to turn a minor loss into a strategic defeat.

What would be the reaction if Iran were to use one of its long-range missiles as a direct-ascent ASAT weapon to knock out a US advanced KH-11 spy satellite? All over the Middle East, Tehran's supporters would be dancing in the streets and handing out candy to celebrate the glorious Muslim victory, while America's allies would fret and try and restrain the US government from any "disproportionate" response, just as they did in 2019 after Iran shot down a US Global Hawk drone. After such an attack, even if the US were to retaliate by launching a two- or three-week bombing campaign, this would leave the Iranian government with greater prestige than ever. In order to gain such a "victory," would it not be possible that Iran would fire off dozens of missiles at US targets in low Earth orbit?

Just because a course of action is irrational in Western terms does not make it so to our enemies. Space has a symbolic value that we, as possessors of spacepower, tend to underestimate. Indeed, one of the persistent arguments against US space weapons is the idea that nations may object to the symbolic idea of being overflown by such systems. However, with or without space weaponization, hostile peoples, states, and organizations will continue to see themselves as being "humiliated" by US and Western space superiority and will seek to neutralize or destroy it wherever and whenever possible.

Opponents such as rogue states and potential foes seek to nullify overhead reconnaissance by the widespread use of underground facilities in North Korea, Iran, and Lebanon, to name only a few. Equipment that can quickly and cheaply dig tunnels for transportation, such as those made by Elon Musk's Boring Company, are making this strategy ever more attractive. Other responses have included decoys and false signatures; these efforts will become ever more sophisticated as more information becomes publicly available about US measurement and signature intelligence prowess.

The 1998 Rumsfeld Commission to Assess the Ballistic Missile Threat to the United States wrote that, "Concealment, denial, and deception efforts by key target countries are intended to delay the discovery of strategically significant activities until well after they have been carried out successfully." This does not always work, but it has been shown to work well enough when combined with political actions so that Iran and North Korea have now obtained a real, if precarious, hold on long-range missile technology and may now be able to build devastating warheads for their missiles.

It must be expected that these foes and others will seek to move from passive or Earth-based anti-space measures to active ones as soon as the political opportunity, combined with the technical capability, presents itself. They may not only be able to use amateur "space watchers" to gather targeting information, but they may soon have access to European space surveillance systems data that, theoretically, will be able to track the so-called US "stealth satellites" (if they exist). Detailed information on US space assets will become more and more widely available in the near future, and enemy space operations planners must assume that they will have even less ability to hide themselves than do their counterparts on the US Navy's aircraft carriers.

If one is looking for surprises, then one place to look might be in the field of directed-energy weapons. Lasers were invented in 1961; the basic technology is openly available. Simple laser pointers, for use in classrooms or on weapons, are for sale everywhere, and Iran claims to have developed its own laser-guided bomb system. High-powered lasers have proven difficult but not impossible for the US to build, but there is no reason to think that if nations such as Iran or North Korea are willing to make the effort (they already have, to develop nuclear weapons), then why wouldn't they do the same for lasers or high-power microwaves? It may also be that certain types of these weapons will be for sale on the international arms market. It has been reliably reported that laser-based IRCMs have been sold to nations such as Saudi Arabia to protect their chief-of-state aircraft. The US and Israel may revive their efforts to build the Mobile Tactical High-Energy Laser (MTHEL), and this project would inspire emulators from France to China.

A laser with enough power to blind an incoming, IR-guided missile would also, in a slightly different configuration, be able to be mounted on a satellite and used to attack another satellite at fairly close (by orbital standards) range. The complex US or Western testing, qualification, and integration procedures probably would not apply to a rogue state's weapon, but there is no reason to believe that a primitive low-powered space-based laser weapon is beyond the reach of many of America's potential enemies. Just because we don't think that such weapons could be developed by Iran or North Korea (or their future equivalents) doesn't mean they lack the imagination or the will to do so.

There is also the possibility that non-state actors such as al-Qaeda could build and use their own anti-satellite weapons. As we have seen in Lebanon, Hezbollah has been able to obtain relatively long-range missiles such as the Iranian Zelzals that Israel was able to destroy on the ground. These seem to be the equivalent of the Soviet FROG-7 or Iraq's Al-Samoud, with a range limited to a hundred kilometers or slightly more. In the future, could a well-financed terror group build a launch vehicle? Could terrorists buy enough parts from North Korea or elsewhere to be able to put together a weapon capable of hitting a target in LEO? Such an attack is certainly possible given enough time, money, and a safe place to operate—preferably near the equator. Al-Qaeda's old stronghold in southern Somalia might be an ideal place from which to launch such a weapon. Targets might not only include US military satellites but also highly symbolic targets, such as the International Space Station (ISS) or one of NASA's future Orion missions.

Again, it must be stressed that America's current and possible future enemies are devoted to the development of new weapons against which the US has not yet developed an adequate answer. On 9/11 the terrorists were able to use the known procedure of aircrew cooperation with hijackers to accomplish their hitherto unthinkable goals. The Arab/Muslim culture is one that has a very high degree of respect for weapons technologies, even if it has not shown much ability, so far, to develop new ones.

Yet throughout history, victory has often gone not to the side with the best new technology, but to the side that makes the best use of what it has available. A prime example was during Israel's War of Independence, when the Jewish side lacked artillery and, in the early stages, lacked conventional airpower and armor. As well as smuggling arms in from Europe, the Israelis improvised and developed an effective arms industry out of the small, clandestine, small-arms workshops that had been created during their underground struggle against the British.

Today we must anticipate that sophisticated, clandestine arms industries are being developed by any number of rogue states and organizations. Could these produce weapons capable of harming US space systems? Leaving aside the obvious possibility of a terrorist using a rocket-propelled grenade to hit an antenna or a firearm to assassinate key personnel, what types of space weapons could be built using widely proliferated technologies and how effective would they be?

The two types of what could be termed improvised space devices (ISDs) mentioned above, the direct-ascent ASAT (DA-ASAT) and the low-powered laser satellite (LPLS) could be built without too much effort by almost any semi-developed nation with access to missile technology. North Korea's development of the Taepodong missile series, while hundreds of thousands of its own citizens starved, is one extreme example of what can be done with technology available on today's open market. Nations with Scud and Scud-derived missiles could, if they choose, begin work to modify these into ISD launchers.

Beyond this, the widespread development of small space launchers, such as the Euro-Italian Vega and the new Pakistani satellite launcher, will result in an ever-expanding number of individuals with the expertise needed to design and build rockets capable of reaching orbit. It is almost inevitable that some of these people will be willing to sell their skills to unsavory regimes. Identifying and tracking these individuals will be even more difficult than tracking individuals involved in nuclear weapons proliferation.

It should be expected that powerful ballistic missiles disguised as space launch vehicles will begin to pop up in some unexpected places, for example in Nicolás Maduro's Venezuela, Vietnam, or Bangladesh. All these nations resent and fear the regional hegemonic power and may imagine that ballistic missiles, whatever they are armed with, will give them a way to deter the US, China, or India.

For obvious reasons, America's enemies will also seek the element of surprise. A DA-ASAT launched from an identifiable launch pad can be recognized and identified in time for its targets to change their orbits enough to avoid it. The same weapon launched from an unexpected site will have precious seconds, possibly minutes, before the US space system controllers decide whether and how to respond. More technologically advanced foes might choose to use fast-burn boosters of the type that Russia claims to be prepared to use to negate space-based, boost-phase missile defense weapons. This would tend to lead toward a requirement for an automated self-defense system analogous to the US Navy's Phalanx anti-missile system for all US major space assets.

Defending against an LPLS may be even more difficult, since one could be launched in the guise of a laser/optical communications experiment by a rogue or hostile nation. Some method or template should be developed to allow the US to differentiate between genuine laser communications satellites and those that have potential to be weapons. There may also be a need to recognize whether and when a satellite is under attack. An LPLS attack on one of our current generation of satellites might only be recognized by the gradual reduction in the power generated by it solar arrays, and this could be attributed to design or manufacturing flaws in the satellite itself.

America was lucky that the right sensors were on board the US satellite that was hit by a Chinese laser. If this event had passed undetected, then the door would have been opened for other, more destructive laser strikes against US satellites. At least now it is possible to identify where an Earth-based laser beam is coming from. The question may arise in the future if a laser attack is carried out from a place like Chad or St. Vincent or Mindanao—places where the local government has neither the resources nor any rational motivation to do so—how will the US or any other space power respond? Are we dealing with a subnational terrorist or criminal group, or is some other state using the territory of a weak neutral to launch an attack against the US? If past experience is any guide, the intelligence available will be confusing and ambiguous.

It has been claimed that America's advanced-imagery satellites already have some shielding against laser attack. The effectiveness of such shielding is unknown; it might be possible that the shielding has never been properly tested. Another countermeasure might be to just move out of the range of an LPLS. The beam from an LPLS would probably not be too damaging at ranges greater than a couple of hundred meters, so a small shift would be enough, at least until the attacking satellite moves again into targeting range. By threatening a valuable satellite, an attacker can force it to maneuver and thus expend fuel, shortening the operational lifespan of the spacecraft.

Another class of space weapon that must be anticipated is the "pseudo system" or pseudo space weapon. These are the equivalent of the Q-ships of World War I, which sailed disguised as civilian ships but carried a powerful set of guns that allowed them to attack and sink almost any enemy that was imprudent enough to come within range. One can speculate that a large Earth sensing satellite such as the European Space Agency's (ESA) Envisat could be launched with a substantial power-generating capability. After launch, it could be announced that one or more of the power-hungry onboard sensors had failed but that the mission would continue. Thus camouflaged, a satellite with an effective directed-energy weapon could remain in orbit for years with no one the wiser. When needed, it would be in a sun-synchronous orbit that could be modified in order to get within range of a US Keyhole-type satellite.

Pseudo debris, co-orbital anti-satellite weapons disguised as space debris or as a disabled or non-functional spacecraft, are another example of this class of weapon. A number of such items could be pre-positioned in various orbits, including those dedicated to parking older satellites that have outlived their useful lives. At the right moment, they could be activated and could perform a real "space Pearl Harbor."

Another type of pseudo space weapon might be a series of orbital maneuvering robots designed to clean up space debris. The orbital maneuvering ability of these combined with grappling arms—perhaps a miniaturized version of the shuttle's—could provide the nation or entity that controls the spacecraft the ability to do serious damage. Indeed, such craft could be used repeatedly to attack target satellites until they run out of fuel or are destroyed.

In the complete absence of active defenses on US satellites, this class of space weapons can only be countered by large-scale and effective space surveillance, including up-close, in-orbit inspection of suspicious objects. Even careful observation of a space object may not be able to detect its true nature; how would one know what to look for? Potential foes could and probably would object that any US craft capable of making such inspections could also be used either as an anti-satellite weapon itself or as a platform for one.

This brings up the important question of dual-use systems. Many of the satellites and probes that are built in the US and elsewhere can be turned into weapons with relatively little modification. The small satellites built by Surrey Satellite Technology in the UK are excellent examples of this. In fact, just about any spacecraft with a navigation system and a thruster could theoretically be turned into a space weapon. This situation may be analogous to the one in the 1930s when it was fairly easy to turn airliners into bombers, as was done with the German Ju 52 and He 111, and the American DC-3 and B-18.

Just as "work expands to fill the time allotted," it may be said that "content expands to fill the bandwidth allotted." While civilian demand for tradition communications satellite service has not grown as much as had been expected, other types of communications services, including digital satellite radio and Ancillary Terrestrial Component/ Mobile-Satellite Service (ATC/MSS) networks are being financed, developed, and built. The capacity of US military communications satellites is already vastly oversubscribed. In spite of the efforts of the acquisition community and the US space industry, there is little chance this situation will change in the near future.

The need for greater and greater degrees of "reachback" capability, particularly for telemedical and intelligence purposes, will ensure that the US will be a major purchaser of commercial satellite communications services for many years to come. The only long-term solution may not be the technology inherent in the TSAT laser/radio frequency (RF) communications satellite program, but in the very large piezo-electric bimorph structures mentioned in Ivan Bekey's *Advanced Space System Concepts and Technologies* (American Institute of Aeronautics & Astronautics, 1986). Such large structures would not only be highly capable communications apertures, but they could be designed to be able to take a hit and keep on working—or, as the saying goes, "to degrade elegantly."

This is in contrast to present and future commercial communications satellites, which are not only unshielded against EMP and directed-energy weapons, but are basically one hit away from total destruction by accidental or purposeful collision. It might be possible for future contracts with the owners of these systems to persuade them to buy shielding and extra maneuvering power for their spacecraft, but the weight penalties are such that the companies involved would demand heavy subsidies.

In the late 1970s, the US Defense Department objected to a proposed arms control agreement on anti-satellite weapons because officials believed the Soviets' "dedicated ASAT weapons would be impossible to verify since the SS-9 booster was used for other missions." They refuted arguments that the Soviets would not use an untested, covertly deployed ASAT system, saying there were "ways to disguise an ASAT test under the cover of activities such as spacecraft docking."

This creates a need to monitor all foreign space programs in order to ensure that they are not disguised weapons development programs. This should be done long before their spacecraft are launched; ideally, it should begin while they are still in the pre-development stage. It should be clear that such monitoring should be carried out discretely. Studies should be undertaken to determine what characteristics and signatures pseudo space weapons would have. An example of what to look for might be excessive capacitor capability or, if we cannot detect that, then estimating weight-to-power ratios in light of the spacecraft's ostensible mission might work.

"Space Situational Awareness" has long been seen as an essential aspect of a nation's spacepower, and the need for ever more refined and capable space surveillance systems is not going away. The ability to hide one's own assets and to detect those of the enemy—and those belonging to neutrals—is just as much a part of space warfare as it is of any other kind. The sensors needed to accomplish this goal in the future will not necessarily be new versions of what has worked in the past.

The US will have to put a lot of effort into making sure that its future satellites are as hard to detect as possible. This not only means that it will have to look for ways to hide them from adversaries, but also from "friends" such as the proposed European Space Surveillance System and from amateur sky watchers. Spacecraft designed to work in LEO will have to be made as small as possible. Size and stealth, rather than weight, may be the most important design priorities of the future. In order to be effectively hidden, reconnaissance satellites may have to fly deep into space, beyond geosynchronous orbit, and radically change their orbital attitude and inclination before returning to LEO. Future US Space Situational Awareness networks will have to be as concerned with concealment of US assets as they are with detecting foreign ones.

It is important to recognize the limits of anyone's, let alone this author's, crystal ball. Giulio Douhet, in his 1921 book, *The Command of the Air*, got just about every one of his many technical and military predictions wrong. Most importantly, he failed to understand that the belligerent powers would refrain from the use of poison gas. Therefore, the truly devastating mass attacks on the great cities of Europe simply did not happen.

While his belief in the decisive nature of strategic airpower was shared by practically all the senior allied air commanders of World War II, the debate goes on and on over just how effective these bombing campaigns really were. The mass panic and collapse of civilian morale that Douhet anticipated never occurred, either in England or Germany, or even in Japan. Would it have been different if the belligerents had used gas, as Douhet and so many others expected? It is impossible to know.

Another important flaw in Douhet's theory was his insistence that the Italian navy "willingly give up its auxiliary aviation in order to increase the Independent Air Force." If Italy had actually equipped itself with an effective naval air arm, the outcome of many of the sea battles in the Mediterranean from 1940 to 1942 might have been different. Spacepower cannot be concentrated or centrally controlled, but, like airpower, the different elements must be made to work together so that, at the very least, the different elements are both deconflicted and defended.

The US and other powers will seek ways to use space weapons to enhance the effectiveness of the ones they already have. Strategic air-power may in the future depend on both kinetic and directed-energy space-to-Earth weapons to suppress enemy air defenses that cannot be destroyed by conventional anti-radar missiles or electronic attack. Very powerful long-range surface-to-air missiles such as the S-300 (SA-10) or S-400-type missiles may be difficult to overcome using cur-rent systems. Israel may have come up with a partial answer with its air-launched Rampage ballistic missiles. Future versions of the S-400 may make it far more difficult to establish air superiority, let alone air supremacy over heavily defended areas. Space-to-Earth weapons could hit and destroy these systems, putting far fewer assets, let alone men and women, at risk than would be the case with a campaign limited to endo-atmospheric weapons systems.

For spacepower theorists, the lesson should be that we cannot anticipate what weapons and support systems will and will not be used in a future conflict. Will, for example, civilian communications satel-lites be immune from all forms of attack? Will GPS, GLONASS, or Galileo only be targets of selective jamming, or will they be hit with kinetic ASATs on the first day of the war? What will be the role of Earth-based lasers? Theory can only provide policy makers with a very rough guide. Navy Commander John Klein, in discussing the maritime strategic theories of Sir Julian Corbett, explains that "Corbett makes it clear that theory and strategic principals are never a substitute for good judgment and experience."[44] The US Space Force should, if we are lucky, develop men and women with these virtues.

Senior political and military commanders will have to make their decisions based on political, technical, and operational reality and not on what is or is not theoretically desirable. War, at its most basic level, has been described as "too few men, carrying too much gear, fighting too many enemies, with too little ammunition, in the worst possible weather, at the fold in the map." At the moment, everything that is truly important about military spacepower should be aimed at easing the burden of those "too few men."

CHAPTER 11

PLAYING NICE WITH OTHERS: FRIENDS AND ALLIES

PART 1: EUROPE

Without question, America's most difficult ally is France. Also without question, France is the one American ally that has taken the concept of spacepower, in all its aspects, more seriously than any of the others. The main reasons for this are, first of all, since de Gaulle, France has been obsessed with safeguarding its strategic autonomy vis-à-vis the US. Second, and, more interestingly, France has a tradition of government-driven technological development that goes back to Louis XIV's multipurpose minister Jean-Baptiste Colbert.

However, the one man most responsible for the French space program—and for the French nuclear program too, for that matter—is Robert McNamara, US secretary of defense under JFK and LBJ. Before Kennedy took over in 1961, France had been working both on missiles and nuclear weapons. But McNamara was deeply hostile to the idea that America's allies would develop independent nuclear forces that would, by their very existence, complicate his finely calculated, delicately balanced (and unworkable) nuclear doctrine. He barely tolerated the British bomb, but his determination to stop France from building its own *force de frappe* only made de Gaulle more determined than ever to build a fully independent nuclear weapons array.

A few years later, the US, still under the baleful influence of the former SecDef's philosophy (as well as the commercial interests of the Intelsat consortium), the US made it almost impossible for France to launch an experimental communications satellite, Symphonie, on an American-made launch vehicle. This pushed the French, supported by a few other Europeans, into building the Ariane program. "*On fera Ariane parce que le Français a besoin de reve*" ("We will build Ariane because the Frenchman needs a dream")[45] is how president Georges Pompidou put it in 1973.

A Parisian friend added, "So the Frenchwoman can go about her business while the Frenchman hangs on to his dream."

France was, and is, unable to pay for its ambitions in space on its own, so it has pressured (and sometimes bribed) other European nations to join various programs. Ariane was, up until Elon Musk's partially reusable Falcon 9 launch vehicle came along, a great example of the success of this policy. France's other space programs have not been as spectacularly successful as Ariane, but most of them have not been failures either.

The question arises, Why is France, alone amongst our allies, taking spacepower seriously as a national goal? The answer involves France's unique history, but it also involves France's difficult strategic geography. France traditionally has been a land power; its armies have been the key to its survival and to its great power status. When they failed, as they did in 1812–15, in 1870, and catastrophically in 1940, the results were disastrous. But France has long and vulnerable coastlines and requires two navies—one for the Atlantic and the other for the Mediterranean. This imposes costs on the French government that its traditional rivals, Great Britain and Germany, can avoid. If it wanted to, the UK could (and often does) cut its land forces to the bone. In the twenty-first century it has done so. Germany can afford to have a tiny navy, mostly oriented toward the Baltic. It sometimes appears as if the German navy largely exists to give Germany's shipbuilders something to do.

France traditionally, however, cannot get along without both an army and a navy, but it also cannot afford to strongly support both. Imagine for a moment what would have happened if France had refused to build its large battle fleet in the 1930s and instead invested in extending the Maginot Line along its border with Belgium.

Today, the answer to France's geostrategic problem is, to put it bluntly, NATO. This fact, however, deeply wounds the national egotism of its ruling class. Occasionally a US president will make an effort to soothe the French *"amour-propre;"* Eisenhower, JFK, Nixon, and George H. W. Bush all tried to accommodate France's need to be at the center of global decision-making, but except for a few limited occasions, they failed. French leaders seem to dream of the seventeenth and early eighteenth centuries when France dominated Europe. The fact that their hegemony was taken from them by England—the original "Anglo-Saxon" power—just makes the power of America—the other great Anglo-Saxon power—even more intolerable.

The desire to match or at least challenge US spacepower drives much of what France is trying to accomplish in space, but it's not the only reason. The French have a tradition of long-term, government-led engineering projects; back in the seventeenth century they built canals, ports, and fortresses. More recently we can see the effort they put into the high-speed TGV trains and, less successfully, the Minitel information network. Building a technologically advanced space infrastructure falls naturally into this historical pattern.

While de Gaulle and his Gaullist successors made limited progress in building up the French space industry, it was the socialist (in name at least) president François Mitterrand who put France firmly on the path to becoming a strong second-level player in the global race for spacepower. He came under the influence of a remarkable Gaullist woman—Marie-France Garaud—who effectively pestered him into taking space seriously. Mitterrand's assistant Jacques Attali recorded in his diary how, in a January 1983 lunch with the president, *"Elle fit une grand passion pour la militarization de l'espace"* ("She had a great passionate for the militarization of space").[46] Mitterrand was incapable of being passionate about anything (except perhaps the opposite sex), but he didn't entirely ignore her advice.

It was under Mitterrand that France began its two major military space programs: the Syracuse military communications system and, most significantly, the Helios series of optical spy satellites, first launched in 1995, which was based on technology first developed for the civilian SPOT commercial remote-sensing satellites. Mitterrand also choose to see Reagan's SDI program as a threat to France's nuclear force and to its technological ambitions. He did his best to challenge it with an ostensibly civilian European research program called Eureka, whose impact, seen from today's perspective, was modest.

As a power with global ambitions (as well as a number of distant overseas territories in the South Pacific, the Caribbean, and elsewhere), France deemed a communications satellite system with worldwide reach to be indispensable. The government began real work on the program in the late 1970s.

The first of three Syracuse satellites was launched in 1984; it was owned by what was then France's national telephone monopoly, France Telecom, which leased transponders to the military. This was followed by Syracuse 2 in 1991 and in 2005 by Syracuse 3, which was the first truly military-grade system with fully secured transmission links, providing France and its leaders with a system comparable to America's 1980s-era Milstars. France plans to follow up with a set of Syracuse 4 spacecraft to begin operations in 2021.

For years, France insisted that the SPOT satellites and their technology were strictly civilian. One can doubt that anyone inside the French government truly believed this, but France's main European partner, Germany, needed this myth in order to hold on to its illusion that "Europe" was somehow a unique zone of peace. France proceeded to use the technology it had developed for SPOT in the successful Helios military imaging satellite program.

Helios was built in collaboration with Spain. It is controlled and the data is analyzed from a headquarters at the old USAF base at Torrejón outside Madrid, which also interprets and disseminates commercial and non-commercial remote-sensing data to the members of the European Union. The European Union Satellite Centre is supposed to support both EU policy making and to support the needs of member states. The center's operations are, it seems, embroiled in the traditional EU bureaucratic fog. The center is not exactly a military organization, but it's not fully civilian either. In this respect, it's a lot like NASA.

Currently, Helios 2A and 2B are operational, and the information from them is shared directly by Belgium, Spain, and Greece. Other EU nations may have access based on bilateral or multilateral deals with France and its partners, but they are not accredited.

While other European nations built their own various Earth observation satellites, which were supposedly for environmental data-gathering purposes (but which provided military and intelligence information and which helped develop their capacity to build more advanced spy satellites), France shouldered the largest share of the burden. The government in Paris assumed that its investments would inevitably lead to French industry winning the biggest share of future EU space programs.

It has not quite worked out that way.

In the case of the Galileo PNT constellation, it was the German firm OHB, based in the northern port of Bremen, that won the contract in 2011 instead of the French. This pattern is often repeated (with Ariane, for example), where most of the propulsion work is now done in Germany.

At the moment, president Emmanuel Macron is so devoted to "Europe" that he is ready to sacrifice French interests in order to preserve his dream. This attitude (one cannot really call it a policy) looks similar to America's Cold War habit of allowing allies to win unbalanced trade deals in order to preserve the health of the alliances. The US economy was strong enough to manage the pain, but the resentments accumulated by 2016 helped elect Donald Trump. For France's pro-EU leaders, one question they need to think about is whether France's economy and society will be strong enough to cope with the industrial sacrifices needed to build up the EU as a new world power.

For France, the permanent problem is that while it can afford to start a series of projects aimed to create French or "European" military spacepower, it cannot afford to fully fund these projects without cutting other parts of its military budget. Consider the situation as it existed in 1991, when the French failed to provide a full light division to the allied forces for Desert Storm and had to be reinforced by a brigade from the US 82nd Airborne. In the previous decades they had paid for a nuclear force, but at the price of reducing their conventional forces to a constabulary aimed mostly at operations in Africa.

In 2017, Macron and his defense minister responded to an aggressive Russian "proximity operation" near a French military satellite with a declaration that France would defend its space assets, and he announced that the French *Armée de l'Air* would now be the *Armée de l'Air et de l'Espace*. The new Air and Space Force was not given any serious budget boost, but the move showed again just how seriously France takes the reality of military spacepower and the possibility of war in space.

When the defense minister revealed the new logo for the French Air and Space Force on September 11, 2020, the head of their space command, General Michel Friedling, said, "Today we see objects that behave in unpredictable ways, that are mobile in space, and which change their orbital planes and altitudes, and which are found in unexpected places; sometimes these objects closely approach other satellites by as much as a few tens of meters." It seems that he was talking about Russian "inspector" satellites.

France may be America's most difficult ally, but in the space domain it stands orders of magnitude above all our other partners. This may make life difficult for the USSF leadership, but if our space generals approach the problem realistically—and without the kind of doctrinal blinders that Robert McNamara wore—then a fruitful and enlightened relationship is possible.

In the longer term, Macron and his successors may find it easier to work with the Americans on serious military space issues than to lead their fellow Europeans in the direction of developing real space-warfighting capabilities. Military leaders in Washington and Paris are going to have to adjust their thinking in some new and uncomfortable ways.

In contrast to France, the United Kingdom is our best ally, the one with which we have a true "special relationship," especially in intelligence matters. Yet in both the military and civilian space areas, Britain's leaders have fallen flat on their faces, at least until recently.

The problem is that the higher ranks of the very influential civil service are, by training and tradition, allergic to anything that is imaginative or futuristic. In the 1960s, Labor prime minister Harold Wilson's talk of the "white heat of the technological revolution" was widely regarded as silly drivel. In the 1980s, the establishment did everything it could to stop Margaret Thatcher from seriously joining Reagan's SDI program or, in the civil sector, from investing in the US space station.

During the 1982 Falklands War, which prime minister Thatcher called "the world's first computer war,"[47] the loss of the destroyer *Sheffield* to an Exocet missile was attributed in part to the crew not being able to use their radar warning gear and their satellite communications transmitter at the same time.[48] At the same time, the US was providing space-based intelligence to the UK. In his memoir, Reagan's secretary defense Casper Weinberger hinted that the NRO had modified the orbit of at least one of its intelligence satellites to support the British. At the time we were still using some imagery satellites that dropped film canisters back to Earth for recovery and analysis. If the US did indeed use one of these precious containers to support the British, it was a costly and important sacrifice. Unfortunately, it seems as if the upper reaches of Whitehall failed to learn this particular lesson.

The leadership in London, both politicians and permanent bureaucrats, imagined that US space assets would always be available to them when they needed them and that they did not need to spend anything more than minimal sums developing their own national systems and technology. A fine example of the kind of free-riding that Trump (and Obama) complained about.

As so often happens, it was left to an independent entrepreneur, Martin Sweeting, now Sir Martin Sweeting, to rescue Britain from irrelevance in space technology. His company, Surrey Satellite Technology (SSTL), was founded in 1985 as an offshoot of the University of Surrey, but it soon evolved into one of the world's outstanding builders of small satellites that could be adapted to multiple missions.

It should be noted that in January 2019, SSTL conducted the EU-funded RemoveDEBRIS mission, which tested a satellite that might, in theory, capture and de-orbit bits of space junk. Naturally, the technology involved could be used in a co-orbital ASAT weapon. If Britain did decide to create an ASAT based on SSTL's expertise, it would be interesting to see if SSTL's corporate owner, Airbus, would acquiesce or would put pressure on the firm to refuse to do the job. In a post-Brexit world, companies like SSTL will probably find themselves stuck between doing things the British government wants them to do and carrying out the policies of their EU-based owners.

Recently, SSTL has been working on a small synthetic-aperture radar (SAR) satellite similar to the Search and Rescue Satellite Aided Tracking (SARSAT) system payload Israel launched on an Indian rocket. If this program, called Oberon, were fully developed, it might give Britain a useful high-resolution, all-weather source of intelligence. The program was supposed to be completed by 2025, with the first experimental model flying in 2022. If the program survives the budget-cutting process, it would probably be subject to delay. If it is successful, it will prove that SSTL is Britain's most important space company.

What finally pushed the British government into increasing its space funding from tiny to small was the issue of "Europe." For Labor prime minister Tony Blair and his pro-EU fellow travelers, increasing funding for the UK's contribution to the ESA was a way to prove that they were "good" Europeans without raising the hackles of the Euro-skeptics. His government set up a new British civil space organization—not very well-funded, but it at least had the virtue of existing and providing a focus for science and exploration activities.

Blair's Tory successors David Cameron and Teresa May did little to change things. In spite of Brexit, they also wanted to be good Europeans, and since the ESA is not formally part of the EU, supporting it on the cheap was as politically cost-free for them as it was for Blair. For the Euro-skeptic Tory leader Boris Johnson, his political calculations on space policy were very different.

First of all, the UK's defense establishment had finally woken up to the strategic importance of space. It had begun testing a set of small, inexpensive imagery satellites called Carbonite in 2018, built by SSTL. This was the first sign that Britain was ready to support the US in space with something other than real estate for ground stations and rhetoric.

The EU decision to exclude Britain from the management of the Galileo PNT may be as big a mistake as the US decision to prevent France from launching its Symphonie satellite. Boris Johnson's government immediately started to look for alternatives. The most obvious solution would have been to build a GPS augmentation system similar to the ones operated by India and Japan, but that would have taken years and would probably have needed more than four or five billion pounds to complete.

Instead, the Tory government surprised everyone by purchasing a half share in a bankrupt internet-in-the-sky firm called OneWeb, which already had sixty-four small satellites in orbit and was hoping to have anywhere between six hundred and nine hundred in its constellation. Now the question is whether the UK and its commercial, Indian-based partner can turn the OneWeb system into a whole new type of PNT service provider with or without US cooperation.

It will take some creative engineering and probably some adroit diplomacy to secure new PNT frequencies from the International Telecommunication Union (ITU), but Britain has traditionally been pretty capable in both these areas. If the British can transform OneWeb into a whole new kind of satellite navigation system, it will be a big step toward the transformation of the traditional space industry away from large, expensive communications satellites in GEO toward swarms of low-cost satellites whose orbits will be mostly in LEO or MEO. And it would be much easier to upgrade with new technology than the current heavy spacecraft designed to provide reliable service over many years. As of now, this is only an opportunity for the UK, and other nations are in the race to build space swarm commercial systems.

Since 1945, successive British governments have not had much luck with their "defence reviews," which, in theory, are supposed to provide policies that adapt the UK's armed forces to the strategic realities of the moment. What usually happens is that these exercises end up forcing the military to do more with less. The 2020 integrated review looks like more of the same. This time, however, the British are taking space (and cyber) into account as an excuse to cut back on their "hard" power forces.

Meanwhile, some of America's other allies are taking baby steps toward building their own space forces. Germany has built itself a rather primitive space-based intelligence-gathering radar. The limits of German space technology should not surprise anyone. Not only does the Federal Republic tend to avoid investing in what it considers "exotic" military systems, but since reunification (and especially since Poland joined NATO), German leaders seem to imagine that all their security problems have been solved.

In military space, Germany is happy to follow France and to let Paris spend the money and take the risks involved in creating genuine "European" spacepower. They may think that, in the end, the French will pay and ultimately Germany will have control. This is the opposite of the situation in the 1970s and '80s when Europe was sometimes described as "a German horse with a French rider."

Germany's civil space programs under the German Aerospace Center (DLR) are slightly less Francocentric and are certainly better funded than its military space programs, but, nevertheless, lack imagination and ambition. Germany is quite happy to be part of the International Space Station partnership but on the whole does not see the global space economy as being a big part of its future. French leaders like Macron may pester Berlin to look beyond its narrow self-interest, but rarely do they succeed in convincing the federal government to take space seriously.

Italy, on the other hand, does take military space seriously. Its COSMO-SkyMed radar and optical satellites were, at the time they were launched in 2003–04, far ahead of anything Germany could build.

Also, as part of a deal with Israel for jet trainers, the Italians agreed to buy an Israeli optical imaging satellite based on the latter's Ofek series. This deal was concluded in 2012.

The story, possibly apocryphal, behind much of Italy's current space policy, goes something like this: In 1983 and '84 after Reagan gave his "Star Wars" speech, the US offered to support some allied research in areas relevant to missile defense. Italian industry wanted access to these contracts, which would have helped it gain a foothold in the emerging world of information technology. The Italian Communist Party, which may not have been in government but was extremely influential, vetoed any participation in Reagan's "mad scheme." Instead, as a consolation prize, Italians were encouraged to join NASA's space station program. Thus, today in Turin, modules and structures for human spaceflight are still being built. Apparently, this may include some for China.

If Italy does become deeply implicated in the Chinese space program, it may find it difficult to cooperate with the US. The Italians were reportedly interested in joining the US Artemis exploration program and hoped to play a role in building the Gateway lunar orbiting space station, but if they are heavily involved with China, it will be hard (or maybe impossible) for NASA to find the political support needed to allow Italy to share in the program.

Italy was also the driving force behind the Arianespace Vega small launch vehicle. The rumor has it that Giovanni "Gianni" Agnelli, long-time head of Fiat, pushed the Italian government to demand that, as the price for supporting future work on the Ariane program, Italy be allowed to take the lead in building a new launch vehicle.

Although Vega failed to orbit a UAE reconnaissance satellite in 2019, the launcher, which uses solid-rocket propulsion for its first three stages, has been a moderate success. But the most important and unsaid aspect of the program is that it gives Italy the technology to build IRBMs and medium-range ballistic missiles (MRBMs) anytime it wants to. Italy's geographic position makes it vulnerable to ballistic missiles deployed in places like Libya, Tunisia, or Algeria. None of these countries are immune from being taken over in the future by hostile, possibly Islamist regimes which might like nothing better than to threaten Rome, headquarters of the infidel Catholic Church, with a missile attack.

Judging by the inability of Western nations to effectively deter Islamist violence, an Italian threat to allow retaliation with its own missiles may not help much, but the government in Rome might imagine it would be better than nothing—especially in the almost total absence of missile defenses.

Since 1945, the Italians' defense policy has been a mess; they depend on NATO, but contribute far less than their fair share. At the same time, they occasionally come up with technology that is both imaginative and cutting-edge, of which the COSMO-SkyMed is a good example.

On October 6, 1973, Israel was surprised by the combined offensive of Egypt and Syria. The trauma of the Yom Kippur War has not been forgotten, and Israel, like the US after Pearl Harbor, is determined never to be caught by surprise again. As soon as they could afford it, and as soon as they had the technology, the Israelis started building and launching their own spy satellites—the Ofek series of imagery satellites, first launched in 1988, and the TecSAR radar satellites, the first of which was launched by an Indian rocket in 2008.

Israel also has launched a number of Amos communications satellites, but due to global oversupply, this model has not been a commercial success.

Israel has a tiny civilian space organization, but its space industry is, for nation its size, amazingly sophisticated. However, Israel may be making the same mistake that many small, spacefaring nations make by failing to concentrate its efforts in a few vital areas. Why, for example, should Israel be one of many nations whose industries are capable of building medium-size communications satellites?

Perhaps Israel could follow the example of Canada, which long ago chose to focus its efforts on radar observation satellites (geared toward keeping track of activities in the Arctic) and on robotic arms (which it built for the US space shuttle, the ISS, and probably for the future lunar orbiting space station called Gateway).

Israel has also built a set of missile defense weapons that are, in many respects, world-class. The Arrow 2 and Arrow 3 interceptors, designed and built with American assistance, have been tested and, for the moment, seem effective. As with many powerful missile defense weapons, it would not take much effort to modify the Arrow 3 into a direct-ascent ASAT.

Does Israel need such a weapon? Iran's attempts to orbit operationally useful satellites have been less than nominal, but it probably will keep trying. The UAE, which has its own space program, has signed a peace deal with Jerusalem and, in any case, it never was a threat. The only real dangers Israel might face from space in the near future would be from Turkey, whose military space program may be in its early stages (but which cannot be ignored) or from Russia or China, which Israel on its own could not do much about.

However, since the US has partly paid for the Arrow 3, it might be interesting to imagine what would happen if the US decided to ask the Israelis to build an ASAT version of the Arrow 3 for the USSF. It would certainly cost less than if the US were to build a direct-ascent ASAT from scratch.

PART 2: ASIA

India is not an official American ally, but since the collapse of the USSR, it's not a hostile nation either.

Years ago in Washington, a representative of the Indian Space Research Organization (ISRO) took me out for a wonderful Indian meal and tried to convince me that their program was 100 percent civilian. He was charming and very sincere, but he failed. I had to tell him that India had one program in particular that scared the hell out of Americans—the IRS.

In fact, the Indian Remote Sensing (IRS) system and its follow-ons are an excellent example of the way that environmental research satellite technology projects can be used to develop military-grade space sensors.

Former Indian president Abdul Kalam was one of those few world leaders who "got" space. He rarely if ever discussed India's military role in space, but he did understand that space exploration and colonization was essential and inevitable, and he insisted that India play a role. He was also a big supporter of space solar-power systems and imagined that India's energy needs would best be met by a large-scale investment in that technology.

If there is one word that typifies the Indian space industry, it is "persistence." No matter what their budgetary limitations and in spite of many setbacks, the Indians have managed to establish themselves as a major spacefaring power. It took years, but they now have a reliable, small payload rocket—the Polar Satellite Launch Vehicle (PSLV). They have done almost as well with the Geosynchronous Satellite Launch Vehicle (GSLV).

In the early years, India stressed that all of its space activities were meant to support the development of the nation. Officials stressed that communications satellites would provide education and telemedical services to India's rural poor. Remote sensing satellites would serve agricultural and environment goals. Naturally, both these technologies had military uses, but for political reasons, not the least of which was India's desire to be seen as a leader of the so-called non-aligned world.

In the late 1990s this began to change. An important turning point was when prime minister Atal Bihari Vajpayee and president Bush agreed in 2001 to allow NASA to support India's first scientific mission to lunar orbit, the Chandrayaan-1 probe. This collaboration opened the way for a wide variety of projects, including the Indian Regional Navigation Satellite System (IRNSS) PNT satellites.

India has been constantly improving the quality of its Cartosat series of remote-sensing satellites; today they can provide India's intelligence services and military with high-quality imagery. India has yet to launch any missile-tracking or early warning satellites, but since the test of a Prithvi missile modified as an ASAT in 2019, it is likely that India will seek to build a full set of spacecraft to support a national missile defense system as well as for its nuclear force.

Faced with China, which is looking more and more like a threat, India will, naturally enough, increase its investments in military space. India not only needs early warning satellites, but it also needs a set of all-weather radar intelligence-gathering satellites. They could buy such spacecraft or the technology for such spacecraft from a number of sources—Canada, Russia, or Israel—but given the way politics works over there, it is more likely that the government in Delhi will choose to develop and build its own radar satellite system, no matter how long it takes or how much it costs.

For many years, Japan's space policy resembled Italy's. For political and constitutional reasons, Japan refused to cooperate with the US in the space national security field and especially in missile defense. Instead, Japan devoted itself to building up an impressive (but ostensibly civilian) space program. Indeed, Japan's space policy sometimes looked like a very expensive way to avoid facing some unpleasant facts about its national vulnerability.

A good example of the way Japan tried to prove that it wanted to help the US (but not in ways that would offend anybody) was its space station effort. Successive Japanese governments were fully committed to the ISS, with the result that Japan's Kibō module is comprehensively equipped for its mission. Compared to the ESA module, Japan's program shows that it refused to skimp on its contribution. Over the years Japan has, by itself, almost matched (and in some cases exceeded) the space exploration and technology-development achievements of ESA with its twenty-two member states.

Japan's H-IIA launch vehicle is expensive and thus not commercially viable, but it does give Japan its own nationally controlled space access system. In theory, it also ensures Japan's mastery of offensive rocket technology, in case it ever needs it. In fact, it proved impossible for Japan to build a space launcher that was both reliable and relatively affordable. In the end, it had to resort to buying parts of the H-IIA from the US. The resulting rocket was still costly but has proved a far better deal than a 100 percent Japanese product would have been.

The now-defunct H-II Transfer Vehicle (HTV) system could deliver six tons of cargo to the ISS at a time. This project allowed the Japanese to develop technologies that could be applicable to building their own manned spacecraft or, with a little imagination, a co-orbital ASAT.

Japan has carried out a series of outstanding planetary science missions. The Hayabusa mission, launched in 2003 to visit the asteroid Itokawa, proved successful and the samples taken by the probe were returned to Earth in 2010. Japan has continued to cooperate with the US and has agreed to play a major role in America's Artemis return-to-the-Moon plan.

Until North Korea lobbed a Taepodong test missile over the Japanese home islands in 1998, Japan was content to avoid building up its own military space systems or to cooperate with the US. The shock that the North Koreans inflicted on the Japanese psyche with this test was amazing. Suddenly the Japanese people demanded that their government "do something."

The government responded with a plan to buy Patriot batteries and SM-3 missile defense systems from the US and to build and launch a set of intelligence-gathering satellites. It built and launched a pair of optical imagery satellites and another pair for radar imagery. Recently the Japanese have been working on a second generation of spy satellites which will go into service within the next few years.

Yet even today, Japan's political system finds it difficult to build any defensive system that is at all controversial. The on-off decision to install a pair of Aegis Ashore missile defense systems is a sign that even when there is an urgent need to defend the home islands, the government finds it almost impossible to surmount even the most trivial of objections—in this case, by some locals who imagined there was a risk of spent boosters from the SM-3 missiles falling on their neighborhoods. So a small group of activists decided that they would rather allow North Korean nuclear weapons to have a free ride to their targets in Japan than to allow their government to make an effort to stop them.

As former president Donald Trump might say, "Sad!"

Until recently, Australia was content to use US space assets for its own purposes, and the US was happy to help. In return, Australia allowed the US to establish support bases for its civil and military space communications, most notably at Pine Gap in the middle of the Outback.

These days, thanks in part to Chinese pressure and partly to pressure from a few Aussie entrepreneurs, a small but promising space industry has begun to emerge Down Under. Along with this is a proposal to establish a new space launch facility on the northern coast. Few nations need space communications and space-based surveillance, sensing, and intelligence gathering more than Australia; fortunately, it is in a good position technologically and politically to take advantage of this potential.

To say that South Korea (Republic of Korea, or ROK) has a highly developed economy is an understatement. So the effort by Obama administration to prevent Seoul from developing its own space launch vehicles seems like the kind of stupid mistake the US made with France in the 1960s and '70s. The excuse that allowing the ROK to build its own rockets would somehow destabilize the peninsula is obviously ridiculous. The North Korean regime has shown zero restraint in this area.

So why did the US government expend so much energy and create so much resentment for so little return? The only reasonable answer is that the State Department's arms control shop is so dedicated to preserving the Missile Technology Control Regime (MTCR) that it is willing to see America's alliances damaged and degraded for the sake of its preservation.

South Korea has been working on a Korea Small Launch Vehicle with the help of Russia. The first two launch attempts of the KSLV failed, but on the third try, they did put a small satellite into orbit. Reports now claim the nations are working on an indigenous, liquid-fueled rocket engine.

In July 2020, South Korea announced that it and the US had agreed to revise bilateral missile guidelines to allow it to use solid fuel for space launch vehicles. The ROK has since been working on a series of solid-fuel missiles that can cover the whole of North Korea from launch sites throughout the South. The US now shows no sign that it values the dogma of the arms control establishment more than it does the strategic interests of America's allies.

One would expect that with its world-class electronics industry, South Korea would excel at building satellites. Its Korea Multi-Purpose Satellite (KOMPSAT) Earth observation series, first launched into LEO by a US Taurus in 1999, is more recently launching into GEO using the Ariane 5.

Even if they may be reluctant to say so publicly, most US allies are now convinced that space is a major theater of military operations. They may not want to give up the idea that space can still somehow be made free of "weaponization," but one can guess that, in their hearts, they know this is out of the question.

Our allies and friends are waiting to see just how serious the US and its new Space Force are about keeping America the dominant power in space. While the USSF leaders can try and make the case that US spacepower is and will remain superior, only the president and Congress—with its budgetary authority—can give a definitive answer.

PART FOUR

THE FUTURE

CHAPTER 12

WAR, ASTROPOLITICS, AND US SPACEPOWER

Everett Dolman's *Astropolitik: Classic Geopolitics in the Space Age* is an essential primer for anyone who wants to understand the basics of space warfare, spacepower, and space politics. Dolman makes the point that, in geopolitical terms, "it is recognized as a fundamental dictum that for any critical power factor a state cannot dominate, its highest priority should be to prevent domination by a potential enemy."[49]

In the nearly two decades since its publication, however, some things have changed; the global environment in which America must operate is far less benign than it was. China has shifted from being a competitive partner to being a strategic rival; Putin's Russia has become, frankly, revanchist; and North Korea has become, at least in a limited sense, a nuclear power—and Iran is close to becoming one. All of these foes are seeking the great advantage to be gained by negating US space systems—using ASATs, electronic warfare, ground-based lasers, cyberattacks, and, most dangerously, EMP weapons.

For eight years the Obama administration allowed US space deterrence to wilt. Perhaps this was inadvertent, or perhaps this was done in the name of international fairness, such as was embodied in its notorious transparency and confidence-building measures (TCBMs). In any case, the need to deter America's potential foes from launching effective attacks on our space systems is urgent. Trump's defense secretary Mark

Esper made the point in a speech in Colorado on August 20, 2020: "The United States is the dominant space power and the gains we possess are being threatened."

During the Cold War, the USSR was effectively deterred from attacking US satellites, even if it did develop and deploy a wide variety of space weapons. The Soviets' inaction was based on the fear that if our space systems were attacked, we would see it as an unmistakable prelude to a massive nuclear strike and react accordingly. America, of course, was likewise deterred from attacking Soviet satellites.

Today, the US reaction to such an attack would be: "So what if they destroyed a US satellite? Nobody was killed. Why is it worth going to war over?" The old responses are just that: old. If, say, China were to shoot down a major US spy satellite in retaliation for, say, US support for the Philippines during a South China Sea crisis, how would an American president react? Would he or she blow up one or more Chinese satellites? Would the US even have such retaliatory capabilities? Would the president authorize an attack on the launch site from which the ASAT was fired? Would he or she consider it an act of war and demand a full-on declaration of war from Congress? How credible would any of these threats be?

Simply put, in the immediate future the political shape of space warfare will be determined by the shape of Earth wars. In turn, the Earth wars will be shaped by political geography—that is, by calculations of national interest.

China and Russia together compose one great Eurasian alliance, although it's not a very secure alliance. However, history shows that alliances between major powers are never really secure, but that does not prevent them from being effective and winning wars. Their central position, dominating what British political geographer Halford Mackinder called the "World Island," would have been a huge strength fifty or a hundred years ago. Today, when the space dimension is taken into account, their geography is neither positive nor negative; their sheer size may be impressive, but what counts is their technological skill and their political will to use it.

The US has, and will probably maintain, its overall technological lead; however, the political will to develop and deploy anything like a "space weapon" or, if required, to actually use such a system, depends entirely on who is sitting in the White House at any given moment in time. In any case, Trump showed that he had been remarkably moderate in his approach to space weaponization. There had not been any crash program to develop American ASATs to match those of China and Russia, nor, in spite of Republican support in Congress, had there been a rush to build space-based missile defense interceptors. Trump got a modest increase in funding for hypersonic and spaceplane R&D, but not much more.

Today, space warfare is a gray area; while US space systems are under constant cyberattack, there seems to be a threshold beyond which America's enemies will not go. What we have learned in the past couple of decades, particularly after Russia's takeover of Crimea, is that gray-area warfare can transition to all-out aggression with amazing speed. In space, if an electromagnetic pulse attack, not generated by a nuclear explosion, could cripple a major US satellite or satellite cluster, would this cross the threshold? How would a president respond?

Of course, this would depend on the president. French President Mitterrand may have spoken better than he knew when he said, "*La dissuasion c'est moi!*" ("The deterrent is me"). He probably was thinking of Louis XIV's line, "*L'etat c'est moi!*" ("The state is me"). In an age when nuclear decision-making is theoretically centralized but in practice may not be (and nobody wants to find out what would happen "in practice"), a head of a state armed with nuclear weapons must consider the likelihood that an effective attack on his or her national space systems is a prelude to an enemy nuclear strike. This was the assumption during the Cold War, but when space systems are under daily cyber threats and often are hit by disabling or degrading attacks with, for example, blinding lasers, it is hard to see where a commander in chief should draw the line.

Having advisers who have studied these questions and developed a set of ideas on how a president should respond is one of the Space Force's most important roles. This could be called Space Deterrence and Warfighting Doctrine, or it could be called Presidential Decision Support. Whatever it is called, it will have to be based not only on up-to-date intelligence but also on the nature of the enemy threat. Without Space Force expertise, a president might get some catastrophic advice from a senior officer who spent his career in the nuclear deterrence field. Mistakes can happen; early warning satellites can give false readings. In a crisis, the men and women in the White House need to have an independent senior space officer at their side as well as generals and admirals who understand nuclear warfighting.

Space has its own rules, for want of a better term, its own "geography." This revolves around gravity and, to a lesser extent, radiation. Both are big factors involved in the twin "gravity wells" that dominate the Earth-Moon system and the numerous small gravitic anomalies, the best known of which are the Lagrange points, including L5 (which gave the old L5 Society its name.) Space warfare will involve using these factors in much the same way as submariners and those who practice undersea warfare have to understand the oceanic environment, including thermal layers and all the various types of salinity, not to mention all the creatures who swim in the world's oceans.

One aspect of the space warfare problem is the need to distinguish between genuine space debris and various space weapons and systems that could be disguised as space junk. The idea of a space mine, a weapon that could be activated when needed, is well known. However, other types of satellites could be disguised as "space junk"; reserve navigation satellites, for example, could be activated after a successful attack on the GPS (or similar) systems. The Air Force has traditionally kept a publicly available catalogue of space debris; this could be used by anyone, including the Russians and Chinese, to help ensure the safety of their satellites. During the Obama administration this changed.

The question of space debris is often used as an excuse not to build or deploy US ASATs. While everyone involved is working on non-destructive methods of disabling enemy satellites, the Russians and Chinese are acting as if the option to physically destroy US spacecraft is viable for them. It is sometimes claimed that this is due to the fact that they depend less on space than do the US and its allies. This may be true, but it is also irrelevant. In a space war, the ability to maintain any sort of space system in the face of a determined attack will be a huge military advantage. Being able to defend operational satellites is a critical capability; it is, however, an expensive one. Why should Chinese and Russian satellites operators not have to worry about US kinetic strikes on their in-orbit systems while Americans constantly have to take into consideration the possibility that at any moment US spacecraft could be attacked?

This one-sided situation is intolerable, and USAF chief of staff General David Goldfein recognized this when, in 2018, he stated that "at some point you've got to hit back." A fine sentiment, but neither Goldfein nor, it seems, the new Space Force leadership are willing to take the steps needed to give America anything like a comprehensive capability to actually fight in space. We can be fairly confident that the US does not have an operational ASAT system and is nowhere near having one; if we did, the news would have leaked long ago.

There are three types of "kinetic" ASATs, and we have reason to believe that China and Russia have got all three. The first and simplest is the direct-ascent ASAT. This is a weapon that launches straight up and collides with its target in LEO. The technology is within the grasp of any power with the ability to build an MRBM or IRBM. There are two types of what are called co-orbital ASATS, the first two described as a Low- and High-Altitude Short-Duration Orbital ASAT Interceptors. Because these interceptor systems "enter a temporary parking orbit, the on-orbit lifespan of these systems is measured in hours, which makes them slightly more complex than direct-ascent weapons."[50] The third type is the Long-Duration Orbital Interceptors (also called space mines). A wide variety of this class of weapons is possible, and they can easily be camouflaged as civilian satellites or space debris. Tracking these weapons is difficult; neutralizing them is, as of now, impossible.

The US is not the only Western nation that sees Chinese and Russian anti-satellite weapons as a threat. The French magazine *Le Point* reported on March 18, 2019, "*La France reflechit a placer des Armes dans l'space, afin de pouvoir exercer son droit a la legitimie defense en cas d'agression*" ("France is thinking about putting weapons in space in order to exercise its legitimate right of self-defense"). The article explains that France may put lasers or other defense systems aboard its satellites, or may build a maneuvering defensive spacecraft comparable to the US X-37B, which, as far as we know, has never been equipped with weapons of any kind. However, the fact that France, which has long paid lip service to the idea of space as an unweaponized sanctuary, is even talking about this is a sign of just how alarming the Chinese and Russian weapons programs have become. As of now, in the kinetic ASAT field, the French seem to have decided not to go beyond rhetoric, though they have the ability to build relatively sophisticated space weapons if they wanted to.

Like the French, the Indians are taking space warfare seriously (and even more so) by carrying out a "kinetic" test of a direct-ascent ASAT in April 2019. Also like the French, they feel it necessary to pay lip service to the idea that space warfare or a so-called "arms race in space" can be prevented by some sort of international agreement. The Indian test introduced a new factor into the US-China military space balance, just as China itself destabilized the US-Russia military space relationship when it conducted its own ASAT test in 2007. The problem is that, unlike the US shootdown of a rogue NRO satellite in 2008, the Indians were unable to devise a targeting solution that did not create a long-lasting cloud of debris.

The biggest change that is on its way is the expansion of humanity into the solar system. Thanks to the ongoing reduction in the cost of getting into orbit, we can already guess at what this will look like in, say, twenty years. In 2040 or 2050, there will be multiple bases on the Moon belonging to the US, China, and possibly Russia, as well as a whole network of small commercial operations doing everything from mining lunar resources to accommodating tourists, researchers, and visiting businessmen. There likely will be a settlement on Mars, with a small "optionally manned" base on the Martian moon Phobos. Commercial prospectors, both human and robotic, will be scurrying around the asteroid belt looking for recoverable assets.

In Earth orbit there will be a wide variety of structures—space stations similar to the ISS (or perhaps made from rebuilt parts of the current station), space hotels, space clinics and old-age homes, a few space solar-power satellites, and lots of space factories using the raw materials from the Moon and the asteroids to build products that are then sent down to Earth. All this will be extremely vulnerable to attack. The US Space Force will find it impossible to avoid spending time and money protecting America's space economy.

The main method of protection will have to be deterrence. Other nations must know that an attack on a US space asset will never go unanswered. Any large-scale war on Earth will inevitably spread into space; the results will be as catastrophic for the space economy as they will be for the earthbound one. It has been argued that the debris from a space war would make spaceflight impossible for hundreds of years. This may be true, but if the global economy does survive, then a serious effort to remove the debris can be expected. In any case, as long as the US remains a superpower, large-scale war, nuclear or otherwise, is not very probable.

What may be more probable is a terrorist attack on a major orbiting facility or an attack by a rogue state disguised as a terrorist attack. Already, direct-ascent ASATs are within the grasp of nations like North Korea and Iran. This capability will proliferate in spite of well-meaning ideas like the MTCR. A group which controls a minimum of territory, such as the Houthis in Yemen or Hezbollah in Lebanon, could acquire an MRBM capable of being transformed into an ASAT and use it. It's not hard to see the motivations for such an attack, even if it brought down extremely violent retaliation on their heads. Imagine how an attack on the ISS would reverberate. It would be a propaganda triumph similar to the one achieved by al-Qaeda on 9/11.

As more and more valuable facilities move into LEO, the more these facilities will be resented and hated as symbols of either Western or capitalist success. Both the Space Force leadership and the intelligence community should keep this nasty aspect of human nature in mind. The reluctance of governmental institutions to imagine what a creative and evil leader or group might do is one of our enduring weaknesses.

In the future—say, in 2040 or 2050—what might a war fought beyond LEO between major spacefaring powers look like? Would it be simply a series of destructive attacks on colonies, mining settlements, and other facilities? Or might it resemble the colonial wars fought between European powers in the seventeenth and eighteenth centuries? Consider the Seven Years War; in America, it is known as the French and Indian War, and some historians call it the first true world war. The goal of the extra-European strategies of the French and the British was not to destroy the other side's colonies and assets, but to capture them and then use them profitably. From an American standpoint, the greatest British achievement was the capture of French Canada, which not only opened up the Great Lakes and the Ohio territory to settlement, but also gave Britain full control of the Hudson Bay fur trade. The British did not destroy French Quebec, but they learned to control and exploit it (which paid off when Canada remained loyal to the crown during the American Revolution).

Or consider the long conflict between the Portuguese and Dutch over control of the spice trade. This struggle largely took place in what is now Indonesia and was eventually won by the Dutch. Both sides wanted to grab the profits the spice trade generated, but neither side had the desire to destroy the spice-producing assets. Indeed, the trade was disaggregated in such a way as to prevent any single attack from wrecking the whole producing and trading network.

In the late twenty-first century it might be possible to build a downhill economy that would be, more or less, immune from destructive strikes, but the mid-century space resource-recovery and manufacturing facilities will be large and will represent huge investments.

So would antagonists fighting over control of the asteroid mining industry or over the Mars settlements choose to destroy them and kill hundreds—maybe thousands—of human space colonists? Or would they try to seize control of these installations and then use them for their own purposes? Could it be that the kind of savage total war which characterized the world wars of the twentieth century (1914–1945) are obsolete, and that wars between major powers will be fought in limited ways for limited objectives? This would certainly be rational, to the extent that any war is ever rational. But on the other hand, we see in the propaganda rolled out in the face of potential or larval conflicts like the ones between Greece and Turkey or between China and India levels of hate comparable to anything we saw in the last century. Such feelings are not easy to square with an effort to control, but not destroy, major enemy assets.

The problem may be that it will be easier to annihilate a space mining facility that has absorbed billions in investments with a strike launched from Earth or elsewhere than it is to actually take control of the installation. It may be hard to imagine a way to physically seize a large, manned space asset without putting the proverbial "boots on the ground." Getting troops into position and supporting them will be very, very difficult and expensive. In thirty years, some future version of Elon Musk's Starship could be turned into a troopship, just as the DC-3 was turned into a paratroop carrier doing World War II. But such a transformation would be much harder to accomplish than the DC-3 to C-47 switch was in 1941. For example, such a spacecraft would need extensive self-protection systems; it also would have to be extraordinarily maneuverable and thus would require large fuel tanks or have the ability to be frequently refueled.

Military leaders in 2040 may tell their political bosses that it would be cheaper to just blow up enemy assets and live with the consequences. It may be reasoned that creating debris clouds in the asteroid belt would be less damaging than creating them in LEO. Cutting off the flow of raw materials to hostile space factories may be more important in the minds of future strategists than grabbing those resources for one's own side. The shape of such a war will probably be driven by the nature of the assets each side has outside Earth's atmosphere. It would be easy to imagine that if China and the US both had large-scale mining and processing facilities in space, neither side would want to be the first to attack them. However, if the war were between, say, China and India, where India had few space assets and China had a lot of them, then India might choose to hit what they saw as major Chinese installations.

If there is a large-scale war in a "downhill economy" context, attacks on distant space assets may not happen in the war's early stages. People, including many realist leaders, tend to believe in the first few weeks of any war that it will be short. It is only when the situation forces them to recognize that the war is going to last a long time that they begin to think that such things as the destruction of enemy economic resources is a good idea.

This logic may help protect large space facilities in the early stages of a future big-power conflict, but as the war drags on, the combatants may choose to strike such facilities.

One expert believes it is more likely that China would try to seize their enemy's space assets and run them profitably, while Russia would probably just destroy them in order to deny their use to anyone, friend or foe.

Fortunately, large-scale war in the twenty-first century is unlikely. Many of the old reasons why major nations fought—resources, global dominance, religious or ideological reasons, or *Lebensraum* (living space)—have little, if any, relevance in our time. Human ability to adapt to new problems is staying ahead of the really big problems. The wars we do see are either about revenge and "honor," such as those in the greater Middle East, or are caused by failed dictators clinging to power, e.g., North Korea and Venezuela. The big potential fight between China and the US is looking more and more like Cold War II, a decades-long struggle for influence and political power with the military threat in the background, but carefully left unused.

ENDNOTES

1   Thomas Karas, *The New High Ground Strategies and Weapons of Space-Age War* (New York: Simon & Schuster, 1983), p. 13.

2   R. Cargill Hall and Jacob Neufeld, eds., *The U.S. Air Force in Space: 1945 to the Twenty-First Century* (Washington, DC: USAF History and Museums Program, 1998), p. 166.

3   Anyone who wants to know the details of the DC-X story should read G. Harry Stine, *Halfway to Anywhere: Achieving America's Destiny in Space* (New York: M. Evans, 1996).

4   "Report of the Commission to Assess United States National Security Space Management and Organization," January 2001, pp. 22-23, http://www.space.gov/commission/report.htm.

5   Kenneth Gantz, ed., *The United States Air Force Report on the Ballistic Missile: Its Technology, Logistics and Strategy* (New York: Doubleday, 1958), p. 22.

6   Charles D. Lutes and Peter Hays, eds., *Toward a Theory of Spacepower: Selected Essays* (Washington, DC: Institute for National Strategic Studies, National Defense University, 2011), chapter 8.

7   Ivan Bekey, *Advanced Space Systems Concepts and Technologies 2010–2030+* (Reston, VA: The Aerospace Press, AIAA, 2003).

8   Joseph H. Ewing, *Twenty-Nine, Let's Go: A History of the 29th Infantry Division in World War II* (Plano, TX: Arcole Publishing, 2018).

9   Boris Chertok, Asif Siddiqi, trans., *Rockets and People* (Washington, DC: NASA History Division, 2005), p. 242.

10   Michael J. Neufeld, *The Rocket and the Reich: Peenemunde and the Coming of the Ballistic Missile Era*, (Cambridge MA: Harvard University Press, 1995).

11 Michael J. Neufeld, *Von Braun: Dreamer of Space, Engineer of War* (New York: Knopf, 2007).

12 Neufeld, *The Rocket and the Reich,* p. 264.

13 Theodore Von Karman w. Lee Edson, *The Wind and Beyond: Theodore von Karman, Pioneer in Aviation and Pathfinder in Space* (Boston: Little Brown, 1967), pp. 267–268.

14 Gantz, *USAF Report on the Ballistic Missile,* p. 26.

15 James R. Killian Jr., *Scientists, Sputnik and Eisenhower: A Memoir of the First Special Assistant to the President for Science and Technology* (Cambridge, MA: MIT Press, 1977), p. 71.

16 Neufeld, *Von Braun,* p. 310.

17 George Pendle, *Strange Angel: The Otherworldly Life of Rocket Scientist John Whitesides Persons* (Orlando, FL: Harcourt, 2005). As an aside, in almost any other country Parsons would have been locked up or ignored; only in the US, or perhaps only in Southern California, could a serious misfit like him be allowed to make the kind of contributions he did.

18 Col. Bruce M. DeBlois, ed., *Beyond the Paths of Heaven: The Emergence of Space Power Thought* (Maxwell AFB, AL: Air University Press, 1999), p. 196.

19 Dwayne Day, John M. Logsdon, and Brian Latell, eds., *Eye on the Sky: The Story of the Corona Spy Satellites* (Washington, DC: Smithsonian Press, 1998), p. 142.

20 Henry Kissinger, *Years of Renewal* (New York: Simon & Schuster, 1999), p. 338.

21 Phyllis Schlafly and Chester Ward, *The Gravediggers* (Alton, IL: Pere Marquette Press, 1964), p. 8.

22 Daniel Graham, *High Frontier: A Strategy for National Survival* (New York: Tor Books, 1983), p. 152.

23 John O'Sullivan, *The President, the Pope and the Prime Minister: Three Who Changed the World* (Washington, DC: Regnery, 2006), p. 215.

24 T. A. Heppenheimer, *The Space Shuttle Decision, NASA's Search for a Reusable Space Vehicle* (Washington, DC: NASA, 1999), p. 412.

25 Donald Rumsfeld, *Known and Unknown: A Memoir* (New York: Sentinel, 2011), p. 274.

26 C. P. Snow, *Science and Government* (Cambridge, MA: Harvard University Press, 1961), pp. 72–73.

27 Arthur C. Clarke, *Greetings, Carbon-Based Bipeds: Collected Essays, 1934–1998* (New York: St. Martin's Press, 1999), p. 22.

28 Rebecca Robbins Raines, *Getting the Message Through: A Branch History of the U.S. Army Signal Corps* (Washington, DC: US Army Center of Military History, 1996), p. 329.

29 John Logsdon et al., eds., *Exploring the Unknown: Selected Documents in the History of the U.S. Civil Space Program, vol. III* (Washington, DC: NASA, 1998), p. 93.

30 Norman Friedman, *Seapower and Space: From the Dawn of the Missile Age to Net-Centric Warfare* (Annapolis, MD: Naval Institute Press, 2000), p. 77.

31 Ibid.

32 US State Department policy paper on Energy, Diplomacy, and Global Issues, August 25, 1965, https://1997-2001.state.gov/about_state/history/vol_xxxiv/h.html.

33 "AU-18 Space Primer," Air Command and Staff College Space Research Electives Seminar, (Maxwell AFB, AL: Air University Press), p. 188. https://www.airuniversity.af.edu/Portals/10/AUPress/Books/AU-18.pdf.

34 Ibid., p. 191.

35 Quoted in "Report of the Action Team on Global Navigation Satellite Systems" (New York: United Nations, 2004), p. 5, http://www.unoosa.org/pdf/publications/st_space_24E.pdf.

36 "2019 Missile Defense Review," (Washington, DC: Office of the US Secretary of Defense), p. xi. https://www.defense.gov/portals/1/interactive/2018/11-2019-missile-defense-review/mdr-fact-sheet-15-jan-2019-updated.pdf.

37 Rosanna Sattler, "Transporting a Legal System for Property Rights from the Earth to the Stars," *Chicago Journal of International Law*, June 1, 2005.

38 Ray Huang, *1587, A Year of No Significance: The Ming Dynasty in Decline* (New Haven, CT: Yale University Press, 1981), p. 52.

39 Quoted in John F. Lehman, *Command of the Seas: Building the 600 Ship Navy* (New York: Scribner's, 1988), p. 409.

40 In Christopher M. Stone, *Reversing the Tao: A Framework for Credible Space Deterrence* (Sun Valley, CA: CreateSpace, 2016), pp. 26–27.

41 Strobe Talbot, trans., *Khrushchev Remembers* London: Sphere Books, 1971), p. 471.

42 "Soviet Strategic Defense Programs" (Washington, DC: US Department of Defense/US Department of State, 1985), p. 16. http://inside-thecoldwar.org/sites/default/files/documents/State%20Soviet%20Strategic%20Defense%20Programs%20October%201985.pdf.

43 Matthew J. Mowthorpe, *The United States Approach to Military Space During the Cold War* (Lanham Seabrook, MD: Lexington Books, 2003).

44 John J. Klein, *Space Warfare: Strategy, Principles and Policy* (Abington-on-Thames, UK: Routledge, 2012).

45 Quoted in Thierry Garcin, *Les Enjeux Stratégiques de L'Espace* (Paris: Bruylant, 2001), p. 108.

46 Jacques Attali, *Verbatin: 1981–1986* (Paris: Fayard, 1993), p. 383.

47 Sandy Woodward w. Patrick Robinson, *One Hundred Days: The Memoirs of the Falklands Battle Group Commander* (Annapolis, MD: Naval Institute Press, 1992), p. xiii.

48 Ibid., p. 13.

49 Everett C. Dolman, *Astropolitik: Classic Geopolitics in the Space Age* (London: Frank Cass, 2002), p. 96.

50 "AU-18: Space Primer," p. 279.